John S. Kingsley

Popular Natural History

A Description of animal life, from the lowest forms up to man - Vol. 2

John S. Kingsley

Popular Natural History
A Description of animal life, from the lowest forms up to man - Vol. 2

ISBN/EAN: 9783337231347

Printed in Europe, USA, Canada, Australia, Japan

Cover: Foto ©berggeist007 / pixelio.de

More available books at **www.hansebooks.com**

POPULAR

NATURAL HISTORY.

A DESCRIPTION OF ANIMAL LIFE, FROM THE LOWEST FORMS UP TO MAN.

BY

J. STERLING KINGSLEY, D.Sc.,

OF THE UNIVERSITY OF NEBRASKA, EDITOR OF THE RIVERSIDE NATURAL
HISTORY, THE AMERICAN NATURALIST, ETC.

Profusely Illustrated.

VOL. II.

BOSTON:
ESTES & LAURIAT.
1890.

CONTENTS TO VOLUME II.

POPULAR NATURAL HISTORY.

NATURAL HISTORY.

REPTILES.

AMONG the living forms the reptiles are distinct, and it is easy to frame a definition which will at once include all its members, and at the same time exclude all that do not belong to the group; but if we take the forms which existed in past times into account, and which are known to us from the fossils in the rocks, the problem is at once extremely complicated. We then find on the one hand forms which intergrade with the birds, while on the other certain forms show a strong relationship with the Batrachia.

SNAKES.

Most degenerate among the reptiles are the serpents, or snakes, animals which inspire the ordinary mind with feelings of horror or disgust. The long, legless body, the expressionless eyes, the cold, clammy feeling, the poisonous qualities of many species, and a remembrance of " the serpent," are all elements which contribute to this revulsion. Yet there is an interesting side to serpent life. Many are interesting in their habits, while many others are among the most beautifully marked objects to be found in nature. In all, the body is covered with scales, and the head with larger plates. Only some of the boas have any traces of limbs. They move with great rapidity by means of the broad scales on the lower surface of the body, each of which acts like a foot, forcing the body along, and then being moved forward to take a new foothold. At irregular intervals snakes shed their skins. At the proper time it splits around the margin of the mouth, and then the snake, by its contortions, or by passing between two sticks or stones, gradually works out of the old integument, leaving it entire, except for the opening at the mouth, but turned inside out. The whole is much like drawing a stocking from the foot, and so thorough is it, that even the thin cuticle covering the eyes is removed with the rest.

A much-disputed question is that as to whether snakes swallow their young. Pages upon pages have been written *pro* and *con*, but as yet one can only give the Scotch verdict, — not proven. The story is often told, that when the snakes are alarmed the young rush into the open mouth of the mother, and seek protection inside her body. It seems as if this question might be easily settled, but as yet no naturalist has ever had satisfactory evidence on this point. Time and time again specimens have been brought to the laboratories, but in almost every instance the contained specimens have belonged to another species, and were undoubtedly swallowed as food, or they were nearly developed young which were in the uterus, a place where they could never get to by crawling into the mouth of the mother.

The question of fascination has also been debated to a great extent, but here the general opinion of naturalists seems to be that snakes really have no power of charming birds or animals, and that the many cases reported rest on erroneous observation. The general tenor of these stories is familiar to all. A quotation from Catesby, a naturalist of a hundred and fifty years ago, who described the animals and plants of the Carolinas, must suffice as a sample. Catesby never saw the operation, but describes it at second-hand as follows: "All agree in the manner of the Process, which is that the Animals, particularly Birds and Squirrels, no sooner spy the Snake than they skip from Spray to Spray, hovering and approaching gradually nearer the Enemy, regardless of any other Danger, but with distracted Gestures and Outcries descend, tho' from the top of the loftiest Trees, to the Mouth of the Snake, who openeth his Jaws, takes them in, and instantly swallows them."

In what could this superstition have had its origin? One cannot say with certainty, but it appears probable that it had its origin in the fact that when a snake invades the nest of a bird, the mother with maternal solicitude stays near the spot, and tries to drive away the intruder with her wings and her cries. Again, it is possible that some animals, like man, may lose their wits upon sudden fright, and hence be for the moment unable to fly from the enemy.

Still, it must be admitted that snakes do exercise a fascination of another sort over human beings, for everywhere we find traces, more or less distinct, of serpent worship. In ancient Egypt the snake and the egg were prominent symbols; among our American Indians the snake held a mystic position, and even in the religion of the Jews it was not entirely ignored, but at times it appeared as an object of genuine worship.

Lowest of all the snakes are some small forms occupying tropical countries, and living almost entirely underground, where they feed upon ants and other insects. But little is known of their habits, and, indeed, specimens but rarely make their way to museums. *Typhlops* is the principal genus.

The next grand division of snakes includes all of the non-venomous forms, as well as a few which are provided with poison-organs. The great majority of them are perfectly harmless, and those occurring in the United States are mostly to be regarded as the friends of man, on account of the immense numbers of vermin they destroy. Some of them feed on insects and worms, while others exhibit a great fondness for rats and mice, as well as the gophers which burrow in the fields of the West. It is true that they occasionally visit the hen-yard and steal the eggs and young chickens or rob the nests of birds, but their damage in this line is inconsiderable. But valuable as these snakes are, no one thinks of encouraging them; on the other hand, all fulfil the old prophecy of Genesis and ' bruise its head.'

The first forms we have to mention are, however, not so harmless, for they embrace the largest of snakes, — the pythons, boas, and anacondas of the tropics. The pythons all belong to the Old World, and have been celebrated from time immemorial for their size. If we are to believe the ancient accounts, the pythons of to-day are sadly degenerated in size, for now no such monsters are found as were formerly described. It is just possible that, as in many travelers' tales, a slight amount of exaggeration was introduced into the old accounts. For instance, in Holland's edition of Pliny's Natural History (1601) we read : —

" Well knowne is it, that Attilus Regulus, Generall under the Romanes, during the warres against the Carthagenians assailed a serpent neere the river Bagradu, which carried in length 120 foot : and before he could conquer him, was driven to discharge upon him arrowes, stones, bullets and such like shot, out of brakes, slings, and other articles of Artillerie, as if he had given the assault to some strong towne of warre. And the proofe of this was to be seene by the markes remaining in his skin and chawes, which, untill the warre of Numantia remained in a temple or conspicuous place of Rome. And this is the more credible, for that wee see in Italie other serpents named Boæ, so big and huge, that in the daies of the Emperour Claudius there was one of them killed in the Vaticane, within the bellie whereof there was found an infant all whole. This serpent liveth at the first of kine's milke, and thereupon taketh the name of Boæ."

In later years the size has diminished, even in travelers' tales; for now the greatest exaggeration does not admit of pythons over forty feet in length, while a rigid application of a foot-rule rarely shows a specimen to be twenty feet long. In the same way fiction plays a part in their swallowing capacities. Many of our popular books depict these animals as killing huge deer and buffalo by enveloping them in their tightening coils. Cold, sober fact dispels this also; for these animals but rarely kill anything larger than a dog, and even should they kill it, their mouth cannot be stretched wide enough to take it in. Still, they can swallow objects of considerable size, for the peculiar structure of their jaws allows them to open far wider than one would believe possible. In man the lower jaw articulates directly with the skull, but in the pythons, as in all reptiles, there are two additional bones between the two, and the effect of this is to still farther widen the gape.

In the New World occur the boas, and the name *boa-constrictor* is familiar to all. There are several species of boas inhabiting tropical America, but the ordinary traveler is oblivious of the differences between them, and they all appear in his note-books as boa-constrictors. In our menageries any large snake, no matter whether it comes from Africa or America, bears this name. The boas are very large snakes, some of them being ten or fifteen feet in length. They live in the dense forests, where they live on the animals that come in their way. They rarely, if ever, attack man, but, on the contrary, are afraid of him. Mr. Bates, for instance, relates his experience with one of them. It was coming towards him, but on seeing him it turned and tried to escape. Our naturalist wished to know more about the specimen — to notice its markings and to ascertain its length, and so began to chase it; but the faster he ran, the faster went the snake, and the result was that it got away from the pursuer.

The anaconda is also a native of the same regions as the boas, but it far exceeds them in size. It is also more bold, and is more or less aquatic in its habits. Mr. Bates must again be called upon to furnish a few facts regarding this species. One night as his boat lay at anchor on the upper Amazon, he was startled by a considerable jar, and in the morning found that his hen-coops which hung over the edge of the water had been broken into from below, and two of his chickens were missing. The natives said that an anaconda which had been lurking about for some time must have done it; but Mr. Bates hardly believed this. In a few days, however, the natives instituted a systematic search for the monster who had made serious inroads on their live-stock. With their boats they explored every place capable of concealing it, and at last their search was rewarded with success. They quickly killed the snake, which measured eighteen feet nine inches in length, and sixteen inches around the largest part of the body. Mr. Bates at a later date measured one twenty-one feet long and two feet in girth, and heard of monsters said to be forty-two feet long.

ANACONDA (*Eunectes murinus*).

These large snakes but rarely venture to attack man, although once in a while they will kill children. Their method is to throw coil after coil of the strong body around the victim, and then to kill it by squeezing it to death. Under this terrible tightening, the bones are quickly broken, and at last the prey is limp and in a suitable condition for swallowing.

FIG. 338. — Boa-constrictor.

The story is usually told that they cover the object with saliva, in order that it may the more easily pass down the throat. This, however, is said not to be the case. It is true that one is occasionally seen beside its prey, the latter all covered with slime; but it is possible that this is the result of having been partially swallowed, and then for some reason disgorged.

The way in which they actually swallow their prey is interesting, even though this feature of their table habits be false. They always begin with the head, so that the hair will not interfere with the operation. Then one of the jaws is moved as far forward as possible, when the sharp teeth take hold of the skin, and thus prevent the object from slipping out while the other jaw is extended in the same manner. Thus the process goes on, sometimes taking hours before it is accomplished, and then the snake creeps away to some protected spot to digest its enormous meal at leisure. All of these large forms lay eggs, and it is said that in some instances they actually incubate them by covering them with the coils of the body.

In our own latitudes we have none of these large snakes; but we do have a large number of forms of much smaller size, some of which offer

Fig. 339. — Chain-snake (*Ophiobolus getulus*).

some points of interest. Among them occurs the familiar chain-snake, — the king-snake of the south. It is really a pretty species, its deep, lustrous black being crossed by a number of narrow yellow or white rings. By the negroes it is regarded with a good deal of respect, for they regard it as the actual king of all snakes. It is especially valuable for its destruction of rattlesnakes, a habit it shares with several other forms. It is swift in its movements, and when it sees a rattler, it makes a dart for its neck, grasps it tightly with its teeth, and then quickly wraps the coils of its body around its victim, and actually crushes it to death. Rattlesnakes know their danger, and strive to escape from this fearless foe.

Further north is a near relative of the form just mentioned, a species which glories in a superabundance of names. Under some one of them, — house-snake, milk-snake, thunder-and-lightning snake, or chicken-snake, — it is known to all. It is bold and fearless, and is fond of lurking about the haunts of man, where it does an immense amount of good in devouring vermin of all sorts. Far more terrific in appearance is the blowing-snake, or hog-nosed viper. When one meets it in the fields, it does not try to escape, but flattens out its head and body so that the former has the most vicious appearance, and strongly recalls one of the poisonous serpents. The animal seems all ready to bite, and nine out of ten will flee before it. Yet it is one of the most harmless forms known. It will rarely strike, and even when it does, it will do no harm; for it has no poison-glands, and its teeth are too weak to do more than scratch the skin. The name blowing-snake refers to the noise the species makes at the mating season as well as when attacked.

But few persons can see beauty in a snake, and yet our common green snake, its skin as green as the grass in which it lives, is really a beauty. It is a most harmless form, and can readily be tamed and handled without its showing the slightest annoyance. Like many another snake it burrows into the ground to pass the winter, and in this act it is extremely social; for where one is found, others may be confidently looked for. Once the writer took part in digging out a number of these from a side-hill, where they had been spending the winter. There seemed to be no end of them. From a square rod no less than three hundred were brought to light, there being sometimes a dozen, sometimes fifty, in a compact bunch just as though they had crawled together for mutual warmth.

The black snakes, or racers, are larger than the green snakes just mentioned, and are far less pleasant creatures on account of their size and strength. They can run with the greatest rapidity, they are good climbers, and they crush their prey in exactly the same way as do the largest pythons and boas of the tropics. They have but little fear, and will sometimes even chase children, as the writer knows of his own experience. It is to the negroes that we must go for superstitious ideas about snakes. They call the common snake the 'doctor-snake,' and assure us that when the rattler kills any other snake, the 'doctor' rubs against the body and immediately brings it to life. Of the nearly allied coach-whip snake, they say that it can whip a man to death.

We have alluded several times to the crushing powers of snakes; and we introduce an account of the pine-snake, or bull-snake, which barely reaches north of 'Mason and Dixon's line,' with Dr. Lockwood's description of the way in which it kills its prey. Due allowance being made for size. it will answer for all. The Doctor put a live rat into the cage occupied by one of these snakes. " Incited by some cause, the rat made a run for the other side of the box. Alas! this movement was the one fatal movement of this little hero's life. In attempting this, it had to cross over a portion of its enemy's body. It was the merest touch. but that touch was death. Instantly every part of the serpent's body flashed into activity, as if the whole had been powder, and a spark of fire had fallen on it. In the merest fraction of a second of time, the reptile that seemed to be lying so languid was transformed into an inverted nest, under which was the poor rat. I looked for the head of the snake. It was under this living nest, holding at the hinder part of its victim, which was doubled up in this strange compression. And stranger still was the wonderful adjustment that half a minute of time served to accomplish. The inverted nest of coils opened at its upper or convex end, like the crater of a miniature volcano. Out of this was evolved the head and front feet of the little rodent, whose dark, lustrous eyes stood out, and neck grew thick from the fearful compression. As the pretty little flesh-colored hands lay upon that fatal upper coil, it did so look like the intercession of helpless suffering with pitiless power. This terrible constrictor, although the act was done in an instant, had fully exhausted all her ingenuity in throwing up this fearful engine of strangulation. It was not merely a series of nest-like constricting coils, but one great coil went transversely over all the others; as when the hand squeezes a lemon, and the other hand is made to help the compression. One could hear the bones crack! . . . Happily, death is almost instantaneous, for it is a literal crushing out of life."

The name bull-snake is derived from its wonderful bellowing note which is much like that of a bull, — loud and without a particle of hiss about it. The reptile fills his body with air, and " then expels it with a bellowing that is really formidable." This snake is extremely active, and when alarmed, if possible, it quickly beats a retreat. If escape is impossible, the bellowing begins, and at the same time the snake emits a most sickening stench, but in what way is uncertain. This smell also serves to call the sexes together, and specimens kept in captivity frequently attract others.

Others of our common snakes abound in damp places, and at times even take to the water, where they swim with great readiness. These water-snakes are frequently regarded as poisonous, but such is not the case. They are utterly without poison-organs, and they do not even wait to kill their prey before they begin to swallow it. A near relative of our common water-snake lives in India, and is remarkable for the ease of mimicry it affords. As we shall see further on, the cobra, one of the most venomous snakes of India. is a very peculiar-looking snake. with its broad head and hood. its peculiar markings. and the like. But this other harmless form simulates almost exactly every appearance of its poisonous relative. It has the same pair of spectacles on the back. can distend the neck in the same way. and in almost every detail of form. habit, and color, the two are alike. Who can doubt but that this resemblance proves of considerable benefit to the innocuous forms? Who. on seeing it. would wait to study the inconspicuous characters which point it out as distinct from the similar poisonous form? ·

The striped snakes, or garter-snakes, are our most abundant forms. They. too. are perfectly harmless, but they are very disagreeable to handle on account of the offensive odor they exhale. After touching one of them the hands retain the smell for a long time, and prolonged washing will hardly suffice to remove it.

A snake. common on the eastern coast of Africa, is most remarkably adapted with reference to its food. It is an arboreal form, spending most of its life in the tops of trees, and wending its way from one branch to another, in its search for the nests of birds. All snakes are fond of eggs, but they are apt to lose a great part of the contents : not so the rachiodon. It takes the egg into the mouth and swallows it entire. It does not reach the stomach in this condition ; for on passing down the throat it comes in contact with a row of saw-like teeth, which cut through the hard shell. allowing the white and yolk to escape. and pass down to the stomach. Not a drop is lost. These teeth are peculiar. in that they are really parts of the backbone. which project through the walls of the throat, and their tips are covered with enamel.

The remaining snakes which we shall mention are all poisonous, but there exists considerable difference in the poison-apparatus in these forms. In those first to be mentioned the poison-fangs are found in the front of the mouth, where they always stand erect and ready for action, like the other teeth. In the other group the fangs. which occur only in the upper jaw, can be depressed so that they lie concealed and out of the way in the roof of the mouth. To the structure of these last we will refer again. In the first group — the forms with permanently erect fangs — these specialized teeth are very sharp. and through them runs a fine channel, which conducts the poison from the gland into the wound. The poison is but a modified saliva. and is secreted in regular salivary glands.

Most celebrated of all these forms is the cobra. or cobra da capello — the cobra with the little cape — of India. It is exceedingly poisonous. and one bitten by it rarely recovers. It is held as an object of superstitious

INDIAN SERPENT-CHARMERS AND COBRAS (*Naga tripudeans*).

regard by the Buddhists. The tenets of their faith forbid the sacrifice of the life of any animal, and this noxious serpent is especially protected by them. In their eyes the killing of any animal is a terrible sin, but to dispose of one of these snakes is a far more heinous crime. On account of this superstition the cobra continues to abound in the Indian Peninsula, and south into the Malay regions. Each year thousands die from its bite, and yet no Buddhist or Brahmin will take the first step to rid the country of the pest. It is estimated that in Hindustan alone the annual mortality from this serpent's bite amounts to over five thousand! Can fanaticism go farther than this?

Let us look a moment at the animal as represented in the plate. It is a large form, remarkable for the 'hood' which it possesses just behind the head. In this region the ribs, instead of being curved downwards as they are in other parts of the body, are straight, and are capable of being placed close against the backbone, or being extended at right angles to it, at the will of the animal. In the latter case it produces this wonderful expansion of the neck, paralleled only in a very few forms. In color these snakes vary greatly, blacks and browns predominating; but most of the specimens are provided with a curious white mark on the back of the hood, which has not inaptly been compared to a pair of eye-glasses or spectacles.

The home of the cobra is in the jungle, where it feeds on lizards, frogs, small mammals, and the like. It sometimes takes to the water, and then it varies its diet with fishes. It is a rather sluggish form, and it knows its powers. It will not flee, but holds its own, with head erect, hood dilated, ready to strike any intruder. The effects of the poison are rapid. In a few moments all parts of the system are affected, and the victim suffers the most excruciating pain, and in most cases death soon intervenes and puts an end to the sufferings. Even where antidotes save the life, there are apt to be periodical returns of the pain.

Notwithstanding the exceedingly venomous character of this serpent, it forms a source of livelihood to large numbers of itinerant snake-charmers. These carry their pets about in peculiarly shaped baskets, and when a crowd is assembled, they give an exhibition. Two or three of the party beat drums and play upon wind-instruments, making a music not over-agreeable to civilized ears, but which seems to exert a species of fascination on these reptiles, which are handled with impunity. In some cases these serpent-charmers are frauds; the snakes they exhibit have been rendered harmless by extracting their fangs. At other times this is not the case, and the performers probably owe their immunity, in a large degree, to an utter absence of any indication of fear. Occasionally, however, one is bitten, and the consequences are the same as with any other person.

The cobra is not the most venomous of serpents. This distinction is enjoyed by another snake from the same regions, which attains a much larger size. Specimens are recorded of twelve and even fourteen feet in length. This species is rare, and has received no common name. Its scientific cognomen is *Ophiophagus*, and is given in allusion to the fact that it largely lives upon other serpents. So poisonous is it that its bite will kill a man in three minutes, an elephant in two hours.

More closely allied to the cobra than the form just mentioned is the asp of Egypt, but it is doubtful whether this is the celebrated species which caused the death of Cleopatra, or whether another form which shares the same common name, and which is figured farther on, should have the honor.

Forms allied to the asp and cobra are found all over tropical Asia and Africa, some extending south into Australia and the adjacent islands. They differ greatly in their poisonous capacities, the bite of some being fatal, while that of others causes but little irritation. In this connection should be mentioned the really poisonous harlequin-snake of our southern states, a black species ringed with yellow and a deep red. It is really a beautiful form, but it is so sluggish in its actions, and so little inclined to bite, that it is usually regarded as harmless.

One group of these poisonous snakes is pre-eminently aquatic. Their tails, instead of being round like those of terrestrial serpents, are flattened into a broad paddle, which is of great use in swimming. Almost all of them live wholly in the sea, the great majority of them being found in the Indian Ocean, while one species extends its range across the broad Pacific to the Bay of Panama and New Zealand.

So wonderfully are these forms modified for their aquatic life, that it is interesting to study their structure more closely. Not only is the tail formed into a paddle, but the body is sharp-edged below, like the belly of a fish. These forms of course breath air; but they can go a long time without breathing, and thus can stay beneath the water for hours. Before going down they fill their large lungs with air, and then the curious little valves on the nostrils close, preventing any entrance of the water. Beneath the surface they swim rapidly, hunting for their prey. When a small fish is seen, they pursue it, strike it with the poison-fangs, and await its death, which follows almost instantly, and is accompanied by an entire relaxation of every muscle. It would be almost impossible for one of these forms, spending its life sometimes hundreds of miles from land, to find a place to deposit its eggs. It is saved all trouble on this score, however, as it brings forth its young — from six to nine at a birth — alive. These young are active creatures, and on their first emergence into the world of waters, they can swim as well, and are as able to take care of themselves as are their mothers. When these sea-snakes are removed from the water, they seem to lose all their powers. It would appear that then their vision becomes indistinct, and they are almost perfectly helpless. Attempts have been made to keep them in aquaria; but the result has not been the one desired. They seemed to lack some essential, and in a few days all were dead.

The remaining poisonous serpents are those already mentioned as possessing fangs capable of erection and depression. When not in use they lie

FIG. 340. — Sea-snake (*Hydrophis*).

folded in the roof of the mouth, where they are concealed by two folds of skin; but when the snake strikes, they are thrown out. The blow, by the intervention of a series of bones, presses the poison-sac, causing a drop or two of the poison to flow down through the hole in the fang. Still, it is not necessary to strike to have the poison flow, for the observation has been repeated many times that the animals can spirt it out to a little distance, without the slightest attempt at a blow.

This snake-poison is rather peculiar in its action. It is harmless, or nearly so, when swallowed, as the digestive juices seem to destroy its poisoning powers. This is a peculiarity of many flesh-poisons, and it is said that the South American Indians, in preparing the curare with which they poison their arrows and darts, test its qualities by the taste. The basis of the curare is of vegetable origin, and is derived from plants of the strychnine family ; but it is said that frequently snake-venom is mingled in the mass. To return to our subject : the poison of snakes needs to be taken into the circulation by means of the lymphatic system. It affects the blood, and through it the whole system, and after death, the blood loses its coagulability, and the body decomposes far more rapidly than under ordinary circumstances. This latter fact is doubtless of value to the snake, as it aids materially in the digestion of the prey.

Various antidotes to snake-poison have been advocated. Possibly the most common is whiskey, or some other alcoholic drink. It cannot be denied that this really has some virtue, but it is not always efficacious. Ammonia, too, taken internally, and also injected into the circulation, is also highly esteemed, and more lately permanganate of potash in a one per cent solution has been injected into the wound, and it is said to be of great value. Besides remedial agents, immediate cauterization, or a cutting out of the flesh around the wound, are of great value. Then, too, the wounded portion should, if possible, be separated from the circulation of the rest of the body by tying it very tightly with a cord, between the wound and the body, thus greatly retarding the spread of the poison, and in this way giving a chance for the remedial agents to work.

Some of the poisonous snakes with erectile fangs inhabit the Old World, and some the New. Of the Old-World forms the various species of vipers are best known. All have a flat, disgusting-looking head ; some live in the tropics, while others venture as far north as the Scandinavian penin- sula and the British Isles. Some are very deadly, while others produce an inflammation scarcely more troublesome than the sting of a wasp, while between these two extremes almost every variation can be found.

Of the few forms we can mention, the first is the horned viper of Egypt and northern Africa generally. As will be seen from our illustration, it is a most disagreeable-looking creature, with a sharp horn or spine above each eye. It lives in the deserts, its brownish white color agreeing with the surrounding sands beneath which it frequently burrows, leaving but the head above the surface. Here it awaits its prey ; but as game is not especially abundant, it often has long fasts. It is said to be very poisonous, and by some it is regarded as the species which terminated Cleopatra's life.

In England the viper, or adder, is the only poisonous serpent, and even there it is not very abundant. Its bite is but rarely, if ever, fatal, but it is very painful and sometimes produces constitutional troubles.

The other snakes to be mentioned all belong to the New World, and from their numbers and their importance they demand more attention than those that have gone before.

Most celebrated of all our snakes are the rattlesnakes, of which the United States contains seventeen distinct species. The principal genus is *Crotalus*, and this derives its name from the Latin. In the language of ancient Rome the castanet worn by dancing-girls was called *crotalum*, and so when Linné was hunting for a name for our serpent, what was more natural than that he should pitch upon this word? For the rattle-

Fig. 341.— Horned viper (*Vipera cerastis*).

snake is provided with a castanet of a most peculiar pattern, and one which is nowhere paralleled in the whole animal kingdom. This rattle terminates the tail, and is composed of a number of dry hollow rings loosely hinged together and ending in a rounded button. Each time the skin is cast, another ring is added to the rattle, and in one case as many as forty-four of these rings have been recorded in the tail of one of these reptiles. Twelve or fourteen is, however, the usual number in our more common eastern species. It is usually supposed that the number of rings in the rattle may be taken as an index of age. This belief is an old one, and

Thomas Morton, in his ' New English Canaan ' of 1632 says, " There is one creeping beast, or creeple (as the name is in Devonshire), that hath a rattle at his tayle that doth discover his age." The idea is, however, like many another existing in common folk-lore, erroneous. The animal may molt several times a year, or the rattles may be lost without leaving any apparent indication of the fact, and in either case any estimate of age would contain an important element of error.

What purpose the rattle plays in the economy of the snake is not easy to see. One hypothesis after another has been advanced, but to each there is a more or less obvious objection. The snake, when alarmed, vibrates its rattle so rapidly that one cannot trace its motions, all that is visible being a haze of light. The crepitating noise produced is not easily described. It has been compared to the song of the cicada, but the resemblance is not perfect.

First among the theories advanced is that the rattle is used as a lure, and that its note attracts insect-eating animals within reach. At first sight this may seem probable, but a little consideration shows that it is not. It supposes an evolutionary process for which there is no adequate cause, and we know that nature does not work in that way. And besides, when hunting, the snake does not sound its alarm. It rather moves stealthily along in perfect quiet, or it lurks in ambush, and waits until its prey ventures within reach.

More probable is the view that the rattle is protective, and that its warning prevents the snake from accidental injury by forms capable of harming it. " The noise itself may not be instinctively fear-inspiring, nor perhaps is the growling of a lion, but in each case experience has taught men and the larger quadrupeds that that growl and this rattle mean not only a willingness to defend, but the certain ability to do deadly harm. This menacing message, clicked from the vibrating tail, has caused many a man to turn back and give the snake a chance to escape."

A third and still more probable view is that the rattle serves as a sexual call, combined with the function of defence just alluded to. It has been noticed time and time again, with these snakes in confinement, that when one springs his rattle, all the others exhibit evidences of their being excited or alarmed,

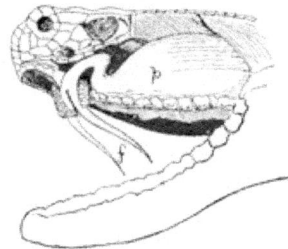

Fig. 342.—Dissected head of a rattlesnake, showing the poison-sac (*p*) and the erectile fangs (*f*).

one can scarcely say which in some instances. In some snakes the sexes are called together by smell, in others by a hissing or blowing noise ; but the rattlesnake has no voice, and he also lacks the strong odor found in some other species.

The other interesting feature connected with the rattlesnake — the poison-fangs — is to be sought at the opposite end of the body from the rattle. We have already alluded to their structure, — the long, tubular fang erected when the animal strikes, the poison-sac, and the apparatus for forcing out the venom, — and hence need not repeat it here.

Our cut represents two of the most common rattlesnakes east of the Mississippi; in the foreground is the more common northeastern form;

Fig. 343. — Rattlesnakes (*Crotalus adamanteus* and *C. horridus*).

while the other, the diamond-rattler, lives farther to the south. The first rarely exceeds five feet in length; the latter reaches eight feet. Our other species need not be mentioned by name, as the differences between them are of such a character as to interest the naturalist alone. All are more or less gregarious, and sometimes immense numbers are found together. Thus at one time, according to Dr. Dekay, two men killed eleven hundred in three days, in Warren County, N.Y., and in other places dens nearly as large have been found in times past. Now, thanks to the war of extermination that has been waged against these reptiles, they are becoming comparatively rare in the more thickly settled regions east of the Mississippi; but on the plains they seem to exist in numbers still unreduced.

The rattlesnakes feed on small birds and mammals which they capture; and in turn they are preyed upon by various forms. Hogs are said to be especially fond of them; their hides and thick layers of fat protecting them to a great degree from the reptiles' venom. When a rattlesnake is alarmed, it instantly throws itself into a close coil, the rattle projecting from the centre, and rapidly vibrating, while the head is held aloft in a threatening manner. It is often said that the snake cannot strike except when in this position. This, however, is not true, although the snake usually assumes this favorite posture before the blow is struck.

As with many another poisonous form, the rattlesnake, in popular estimation, is endowed with various remedial qualities. Its poison even enters into the pharmacopœia of homœopathic physicians. With this latter we have, however, nothing to do; but some of the common superstitions are of interest. The oil is highly esteemed as a remedy in cases of fever and rheumatism; and, says Mr. Ingersoll, in the course of an interesting paper on these animals, "Every summer to this day [1883], the citizens of Portland, Conn., go out to the Rattlesnake Ledges and catch the reptiles with gaff-hooks, the local druggists paying them four dollars an ounce for the oil, which finds a ready sale." Among the other uses to be mentioned are the following: The cast skin and the fat are in repute for curing the bite, — a truly homœopathic doctrine, — while the rattles are believed by the Indians as well as many whites to be of value in parturition, and the dried flesh in curing cases of consumption. There really seems to be but little doubt on one point, — that the flesh is good to eat.

Another erroneous idea — it can hardly be called a superstition — is that which associates prairie-dogs, owls, and rattlesnakes together in a happy family. The story goes that these three forms live together in the most amiable way in the burrows of the prairie-dogs, — but stop! Dr. Coues has put the case far better than I can ever do it. He says that "no little pure bosh is in type respecting the harmonious and confidential relations imagined to subsist between the trio, which, like the 'happy family' of Barnum, had Utopian existences. According to the dense bathos of such nursery tales in this underground Elysium, the snakes give their rattles to the puppies to play with, the old dogs cuddle the owlets, and farm out their own litters to the grave and careful birds; when an owl and a dog come home paw-in-wing, they are often mistaken by their respective progeny, the little dogs nosing the owls in search of the maternal font, and the old dogs left to wonder why the baby owls will not nurse. It is a pity to spoil a good story for the sake of a few facts, but, as the

case stands. it would be well for the Society for the Prevention of Cruelty
to Animals to take it up.

"First. as to the reptiles, it may be observed that they are like other
rattlesnakes, — dangerous, venomous creatures; they have no business in
the burrows, and are after no good when they do enter. They wriggle into
the holes, partly because there is no other place to crawl into on the bare,
flat plain, and partly in search of owls' eggs, owlets, and puppies to eat.
Next, the owls themselves are simply attracted to the villages of the
prairie-dogs as the most convenient places for shelter and nidification,

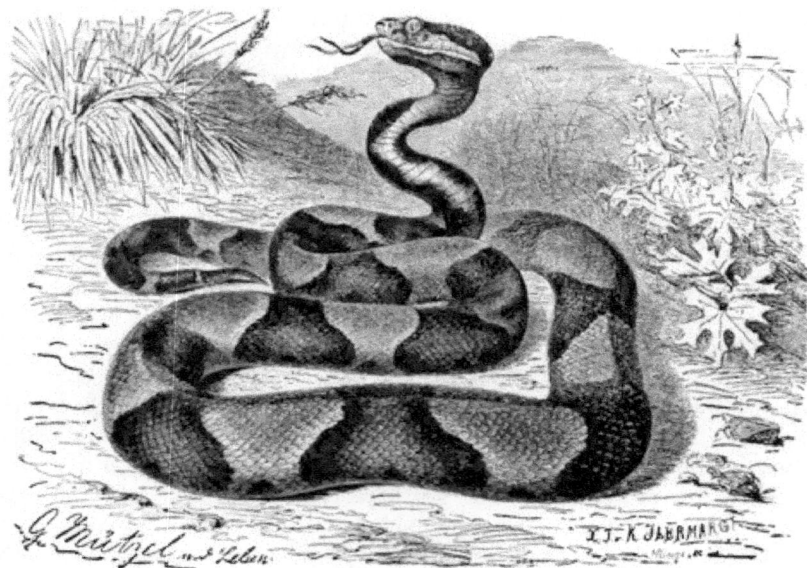

FIG. 344. — Copperhead (*Ancistrodon contortrix*).

where they find eligible ready-made burrows, and are spared the trouble
of digging for themselves."

Others of the poisonous reptiles of America lack the rattle, and in the
tropics the most deadly of these is the celebrated fer-de-lance. This is
a native of Surinam and Brazil. but it has obtained entrance to some of
the West India Islands. In Martinique it has become an insufferable nuisance, or more than a nuisance. It is a quick, aggressive creature, and
strikes without any warning.

In our southern states occurs the water-moccasin and the copperhead,
the latter extending north into New England. Both are poisonous, but the

moccasin far more so than the copperhead. It even exceeds the rattle-snakes in its dangerous qualities; for the latter always gives warning, but the moccasin and the copperhead never do. In the habits of these two there is considerable difference. Though the copperhead is occasionally seen in damp places. it seems to prefer the high and dry land. especially that of mountains. The moccasin. on the other hand. lives in damp places. and frequently takes to the water for its food of frogs. tadpoles. and fishes.

LIZARDS.

The distinctions between the lizards and the serpents and turtles are quite marked. if we take only the typical forms into consideration. Lizards are scaly reptiles with four legs. Snakes, as we have seen, lack the legs. while turtles have a hardened shell. This distinction. however. does not hold good in all cases, for, as we shall see in a short time. there are lizards without legs and scales. In such cases we have to rely on features of internal structure to solve the question where these forms are to be placed. There is besides another group of living reptiles. — the crocodiles and alli-gators. — which cannot be separated from the lizards by the superficial characters given. In this case. too. it is internal structure that must be relied upon.

First among the lizards come the geckos. curious forms with large toes. especially adapted for climbing. Each toe terminates with an adhesive disc. and by means of this they are able to climb up a smooth. perpendicu-lar surface. like the walls of a house. All of them are tropical and semi-tropical, only one species occurring in the United States. and that one is very rare in Florida. In some parts of the tropics they fairly swarm. some of the species living about houses. while others prefer to keep away from the haunts of men. Some species become semi-domesticated. and shortly after sunset they will appear about the houses, utterly regardless of the presence of its rightful inhabitants, running up and down the walls. and creeping into all of the corners in their search for the insects on which they feed. No one hinders them; for all know how much good they do in destroying cockroaches. ants, centipedes, scorpions. and the thousand other pests which contribute to make life a burden in those climes.

Some of them utter a cry, and it is from this that the name of some of the species is derived. a name which has been extended to all. Others croak more like a frog, and still others have other notes. In many the tail is very brittle; and since these animals quarrel much among them-selves, accidents to this member are common. it sometimes being broken completely off. This, however, is a matter of but little importance; for old Dame Nature soon repairs the damage, and in a short time the unfortunate is provided with a new tail as good as the one lost. The geckos have even been seen to eat their own tails. Why, no one knows. It may be that they think them a superfluity, or perhaps they have a peculiar, if not a praiseworthy affection for their own flesh and blood.

Fig. 345. — Geckos (*Platydactylus*).

Sir Emerson Tennent, in his account of Ceylon, writes: "In the boudoir where the ladies of my family spent their evenings, one of these familiar and amusing little creatures had its hiding-place behind a gilt picture-frame. Punctually, as the candles were lighted, it made its appearance on the wall, to be fed with its accustomed crumb; and if neglected, it reiterated its sharp, quick call of *chic, chic, chit*, till attended to."

Among the batrachians we had occasion to mention and figure a peculiar 'flying-frog' from the East Indies. It was a member of a group of tree-frogs, perfectly adapted for an arboreal life. Among the geckos, also climbing animals, occurs a form somewhat similar in that it is capable of sailing (not really flying) in its descent from the trees. In the flying-frog it was the feet that furnished the parachutes; but the flying-gecko (a native of India) combines this with a broad, flat membrane, much like that of a flying-squirrel, except that it extends from the sides of the head and tail, as well as from the body between the legs. When the animal wishes to take its flight, it spreads its legs and extends the membrane, and then sails gracefully downward, much like a flying-squirrel.

Somewhat similar are the flying-dragons of the East Indies, of which there are several species. In these the membranes are more confined to the sides of the body, and when extended are supported by the ribs. These forms are seven or eight inches long, and are beautifully mottled with brown and gray, so that as they crawl over the trunk of a tree in their search for insects, they can hardly be distinguished from the bark on which they rest. Their leaping powers are considerable, though they are excelled by those of the flying-gecko, already mentioned.

What can be more repulsive than the subject of our next cut, the moloch? Its scientific name, *Moloch horridus*, is well deserved. Everywhere it is covered with spines of enormous size, its body is fat and ugly-shaped, and its legs are too weak to keep the animal from the ground; while on the back of the neck is a hump which the 'hunchback of Notre Dame' might have envied. Yet ugly as it looks, this creature is perfectly harmless. All that it desires is to be let alone, and no doubt its extensive spiny armor protects it from many a form which might otherwise be tempted to make a meal from it.

In Madagascar dwells a lizard which must not be neglected. It is the famocuntrata of the natives, who hold it in the utmost detestation, notwithstanding the fact that it is perfectly harmless. From the sides of the head, body, and tail depend a series of fringes contributing somewhat to its appearance, while its bold habits, rushing as it does, with open mouth, towards any one who disturbs it, render it an object of terror. Many are the tales told of its awful habits, — none of them exactly true, and many wholly false, — but none of them exceeds the belief, firmly implanted in every Malagassy breast, that whenever it has a chance it attaches itself to the human breast so closely that it cannot be pulled off, but must be cut away with a knife, or else the victim will surely die.

In tropical America occur the iguanas, some of them of large size, while others are much more moderate in their dimensions. The typical form is shown in our plate. It is a loathsome-looking creature, about five feet long, its wrinkled, scaly skin looking as if too big for the body, while the crest of long, spine-like scales upon the back, and the enormous dewlap under the throat do not add to the animal's attractiveness. Yet, ugly as it appears, it is highly esteemed as an article of food in the regions where it occurs, and all travelers add their testimony as to its palatability. Thus Père Labat, a missionary two hundred years ago, discourses concerning this reptile: "We were attended by a negro, who carried a long rod, at

FIG. 346. — Moloch (*Moloch horridus*).

one end of which was a piece of whip-cord with a running knot. After beating about the bushes for some time, the negro discovered our game basking in the sun on the dry limb of a tree. Hereupon he began whistling with all his might, to which the guana was wonderfully attentive, stretching out his neck and turning his head, as if to enjoy it more fully. The negro now approached, still whistling, advanced this rod gently, and began tickling with it the sides and throat of the guana, who seemed mightily pleased with the operation, for he turned on his back and stretched himself out, like a cat before the fire, and at length fell asleep, which the

IGUANA (*Iguana tuberculata*).

negro perceiving, dextrously slipped a noose over his head, and with a jerk brought him to the ground. And good sport it afforded to see the creature swell like a turkey-cock to find himself entrapped. We caught more in the same way, and kept one alive seven or eight days; but it grieved me to the heart to find that he thereby lost much delicious fat."

The iguana lives in the forests, and feeds, it is said, solely upon vegetable substances, a fact to which it owes its white and tender flesh. It lays about forty eggs in a hollow tree. These eggs are a long oval, about an inch in length, and are covered with a tough, flexible shell. The natives of the Brazilian forests are very fond of them, but unless their taste is disguised by condiments they have no attractions for the European palate, for they have a very oily, disagreeable taste.

Besides this great iguana there are several other species occupying the forests of Central and South America and the West India Islands. Two somewhat closely related species are found in the Galopagos Islands, that locality strangest in all America in its animal inhabitants. These two forms, although belonging to the same genus, are widely distinct in their habits: for while one is almost marine in its life, and is found only on the shore, where it feeds on the sea-weeds, the other never ventures near the salt water, but prefers the thickets of cactus and acacia of the interior. In its habits it is much like the prairie-dogs, digging shallow burrows in the parched soil; and in these they live and lay their eggs. The aquatic form, on the other hand, is a strong swimmer, and can stay beneath the surface of the water for an hour or more.

Long before America was discovered, every treatise on unnatural history had its account of the dreadful basilisk, and each writer did his best to surpass the description of his predecessor. The upas-tree in all its glory did not begin to be as deadly as was this reptile. Even the glance of its eye was death. "This poison infecteth the air, and the air so infected killeth all living things, and likewise all green things," says a mediæval naturalist; " it burneth up the grass whereupon it goeth or creepeth, and the fouls of the air fall down dead when they come near his den or lodging." So terrible an animal must need have a strange origin. No normal process of development would account for the existence of such a beast. So the story ran that it was hatched from eggs over which a snake or a toad had brooded.

When America was discovered, the early explorers found in the forests of South America and Mexico a curious-looking reptile, which had one point in common with the basilisk of fiction, and so this form at once received the name, and upon it was at once disposed the whole stock of old wives' tales. The basilisk of fiction was the king of reptiles, and wore upon his head a crown; the basilisk of fact had a curious lobe on the top of its head, therefore the two were identical. The general appearance of the animal is shown in our cut. It only needs to be said that the animal lives in trees and bushes, and is utterly destitute of any noxious qualities.

The American chameleon, shown in our next cut, is not a chameleon at all. Less does it deserve the other name often applied to it of 'scorpion'; for

FIG. 347. — Basilisk (*Basiliscus mitratus*).

it is in nowise, either in appearance, habits, or poisonous qualities, related to the true scorpions. It shares, however, one feature of the true chameleon, — the power of changing its color. Dr. Lockwood has described this in a charming way. "Now," he says, "begins that wonderful play of colors. It appears first in the normal bronze-brown of the back. Literally they are lively colors; such are the moving changes as the folds of the skin, especially as those on the neck, catch and glance the sunlight. That deep amber is now mellowing into a yellowish brown. A minute more, and it has a bronze coppery tint. Now it runs into an olive-green; anon, a leek-green; at last a pale but bright pea-green. Through all this color-transformation, on the back there is a medial line, extending from the head to the tail, which is always of a paler hue than all the rest. As to the under parts, the customary ashiness is all gone. It is white; but such a white! not glaring, but soft. In fact, I think the tiny scales are now set a little on edge, thus giving the white the aspect of frosted silver. The back, as was said, is green; but now I observe what I have very

FIG. 348. — American chameleon (*Anolis principalis*).

seldom seen, that, so to speak, over this green is a bloom, so that it looks like a frosted green. It is observable that the top of the flat head doggedly retains its dark normal brown. As to the eyelids in this matter of color, I think they are the most to be admired. Each of these little brilliant orbs in constant motion is a perpetual twinkle. In ordinary repose the eyelids are a pretty, pale brown. But these organs are especially susceptible of color-change. Not only will they run rapidly through the whole scale, but the positive colors will be spread in such decided and rapid contrast that it seems as if the order were set to the key-note of a humor which · is alone high fantastical.' These winking eyelids emulate the gems. Now a palish brown, they are smoky topazes. Instantly they become green emeralds, and quicker than one can write flash into the peculiar blue of the turquoise."

This chameleon — for so custom compels us to call it — is a small form with a body rarely four inches in length, and with a tail which stretches out behind to twice that distance. It is a most familiar animal all through the southern states as far north as South Carolina. It makes a most engaging pet, so quick is it in its motions, and so wonderful are its color-changes. The general features can be seen from the cut, the color-changes have already been described; but there remains to be mentioned the dewlap under the throat, which the animal is capable of inflating to a considerable extent when excited. In its color this offers a most striking contrast to the rest of the body, for it is either a rich orange, or more frequently a vivid scarlet.

The chameleons live in the forests and shrubbery, and insects form the bulk of their diet. True chameleons, as we shall soon see, have a long, adhesive tongue, which can be thrust out like that of a toad to capture insects for food; but not so this form. It depends solely on the quickness of its own motions. A few stealthy steps, and then a sudden jump, and the fly is safe within the mouth. These insect-devouring habits render these animals of great value, and in their native country they are protected. It has, however, says Dr. Shufeldt, "an uncompromising enemy in the domestic cat. This animal, I have been informed on undoubted authority, will, when the opportunity presents itself, pass anything, — meat, birds, and even fish, — if there is the slightest chance of securing one of these lizards, of which they seem to be so inordinately fond. The cat will stalk one, just as we have seen them attack some unsuspecting sparrow. Should the lizard be on the trunk of a tree, and low down near the ground, and the cat miss it in her spring, she will frequently, in her disappointment, chase it up the tree, where, of course, the reptile wins in such an unequal race."

There are numerous other lizards in the warmer portions of the United States which present a more or less close similarity to the form just mentioned. Some, however, are much different in their general appearance, but none excel the 'horned toad' in this respect. Horned and spined they are, toads they are not; but still it is useless to protest against popular nomenclature, and so horned toads they will in all probability remain until the end of the chapter. All of them are confined in their distribution to the great plains of the West, and thence south into the tropics. Their colors, various shades of brown, are well adapted to conceal them from the observation on the arid spots which they frequent, and their power of simulating death is another element in their security.

In confinement they make rather interesting pets, and soon become acquainted with their owners, and readily take flies from the fingers. A marked peculiarity is their fondness for a gentle scratching or tickling. They express their gratification in the most marked way; they spread themselves out, or blow themselves up, until they resemble the frog that imagined himself as big as an ox. They feed upon insects, and are especially fond of ants. They bring forth their young alive, the number being, in the north, from seven to eight at a birth; but in the tropics they are more prolific, the number being from twenty-five to twenty-seven. The young are exceedingly active from the first, and can run about as quickly

FIG. 549.—Horned toad (*Phrynosoma orbiculare*).

and as well as their parents. There are several species of horned toads, but in the average museum one will usually find them all labelled *Phrynosoma cornutum*, utterly regardless of the distinctions between the various species.

For a long time the inhabitants of our southwestern territories and the adjacent portions of Mexico have related wonderful stories of the poisonous character of the large lizard, known as the Gila monster. Scientific men, however, have been incredulous, for no other lizard was known to be venomous. At last the question was settled in a very conclusive manner. Dr. R. W. Shufeldt was handling one of these lizards, in the reptile room of the National Museum, at Washington. He had heard of the poisonous character of the animal, and was very careful; but at an unfortunate moment his hand slipped, and the creature bit his right thumb. The rest of the story is given in his own words. "By suction with my mouth I drew not a little blood from the wound; but the bleeding soon ceased entirely, to be followed, in a few moments, by very severe shooting pains up my arm and down the corresponding side. The severity of these pains was so unexpected that, added to the nervous shock already experienced, no doubt, and a rapid swelling of the parts that now set in, caused me to become so faint as to fall, and Dr. Gills' study was reached with no little difficulty. The action of the skin was greatly increased, and the perspiration flowed profusely. A small quantity of whiskey was administered.

Fig. 350. — Gila monster (*Heloderma horridum*).

This is about a fair statement of the immediate symptoms; the same night the pain allowed of no rest, although the hand was kept in ice and laudanum, but the swelling was confined to this member alone, not passing beyond the wrist. Next morning this was considerably reduced, and further reduction was assisted by the use of a lead-water wash." The wound healed in a few days. This experience led to further trials, and with poison obtained from the animals a pigeon was killed in seven minutes, and a rabbit in one minute and a half. The poison is unlike that of serpents in being alkaline, and also in its physiological effects, as it affects the heart and nervous system instead of the respiratory organs.

The Gila monster is our largest lizard, reaching a length of three feet and over. Its body is covered with tuberculated scales, and its colors are black and orange. It lives in Mexico and crosses the boundary line into New Mexico and Arizona. Its large size and peculiar appearance make it a rather common inhabitant of museums and zoölogical gardens.

FIG. 351. — Monitor (*Varanus niloticus*).

The monitors, which inhabit the eastern hemisphere, differ considerably from the forms so far mentioned, in that they live semi-aquatic lives. The typical species, the monitor of the Nile, is a celebrated creature, on account of the amount of folk-lore that has accumulated around it; the principal feature of which is that it acts, as its name implies, as the monitor of the crocodile. In reality it is a really beneficial animal, for its favorite diet is crocodile eggs, and in this way it aids to no inconsiderable extent in keeping down these monsters. It further preys upon the young which hatch from the eggs that escape its search.

Other species occur in the east, some attaining a considerable size. Among these is the water-lizard of India, four feet long, which spends a considerable part of its time in the water, but still wanders about considerably in the lowlands, burrowing about in its search for insects, and especially ants. Larger and still more aquatic is the huge lizard of Ceylon, which reaches a length of six feet, and which is the subject of much superstition and fear on the part of the Cinghalese. It is a strong animal, and, according to Haeckel, a blow from its powerful tail, clad in plate armor, often inflicts a dangerous wound, or even breaks a man's leg.

We must not forget our little striped lizard of the eastern United States, a pretty brown species, marked on the back with six longitudinal yellow lines; beneath it is a silvery blue. It but rarely occurs north of Virginia, but south and west of that state it is abundant, making its appearance in a timid manner about the close of the day. It reaches a length, tail included, of about ten inches.

There are a number of lizards which differ greatly in their structure, but which, nevertheless, we may be permitted to group together on account of a most marked external feature. In these forms we notice first a tendency to a disappearance of the limbs. In the skinks this is not very marked; but in the other forms we notice more and more of a disproportion between body and limbs, which reaches its extreme in the Amphisbænas.

Most of the skinks are tropical, lazy forms, whose chief delight is to bask all day in the sun. At times, however, they start up, impelled by hunger, and hunt for insects and other small animals for food. To-day there is but little interest concerning them; not so a few hundred years ago. The time once was when the more disgusting an animal, or indeed, any object, was, the more highly it was esteemed as a remedial agent; and in those days the dried and powdered skink was deemed a sure specific for a long list of human ailments. To some of the skinks, on the other hand, most diabolical powers were attributed, and in their qualities they were scarcely surpassed by the basilisk.

We have in the United States about a dozen skinks, but to these there has never been attributed any such qualities as to those of the Old World; although of course there are persons who regard them as poisonous, just as they do any and every other reptile.

The next form to be mentioned in this descending scale is the curious Mexican Chirotes, which in everything resembles a snake, except that just behind the small inconspicuous head is a pair of five-toed feet. In the blind-worm of Europe, and the other forms to be mentioned, there is no trace of any legs to be seen from the outside. The blind-worm looks almost exactly like a snake; it is common in Europe, but belies its name in not being blind. It is replaced in the southern United States by a very interesting species, the glass-snake. Allusion was made a few pages back to the habit of the geckos of breaking off their tails; but in the glass-snake this power is far greater. The slightest blow is sufficient to break the tail from the body; but this is an affair of comparatively little importance. In common belief, the lizard backs up to its tail after the disturbance is over,

Fig. 352. — Skink (*Scincus officinalis*).

and then the member quickly grows on again. This is really not the case. What does occur is that the body rapidly grows a new tail, as useful and as fragile as the old one. It occasionally occurs with these, as with other forms with similar capacities of reparation, that the tail will not be entirely detached; then the new tail grows out beside the old one, and a monstrous two-tailed specimen is the result.

This capacity is evidently of considerable good to the species; for the loss of a tail is far better than to lose the life, as might otherwise happen when a hawk or an owl pounces down upon one of these creatures. The glass-snake is yellowish green, varied with black above, and yellow beneath. It spends most of its life in burrowing beneath the surface of the ground, moving about only by the contortions of the body.

Among the strangest of the lizards are the Amphisbænas, the name of which means walking in either direction. To a casual glance both ends of the animal are alike, and in their motions the head seems to have but little preference over the tail, as they wiggle in either direction with almost equal facility. These facts have given rise to the fable that they have two heads, one at either end of the body. In South America they live habitually in the subterranean burrows of the saüba ant, and the natives call them 'mai das saübas,' — mother of the saübas. "They say that the ants treat it with great affection, and that if the snake be taken away from the nest, the saübas will at once forsake the spot." This probably partakes of the nature of a myth; and although the relation which exists between the ants and the Amphisbæna is not yet certainly known, there is considerable probability that the association is of considerable

Fig. 333. — Amphisbæna.

advantage to the lizard; for, on dissection, remains of ants have been found in their stomachs.

In the quotation given in the preceding paragraph this animal is called a 'snake,' and a snake it is in appearance, but not in structure. It has no legs, its body is of nearly the same size throughout; but in all the features of its internal anatomy it is seen to agree thoroughly with the lizards. It reaches a length of about a foot.

These forms lead us in a line of degeneration. There is no doubt but that these slow-worms and Amphisbænas have descended from more normal lizards, and that the whole tendency of their development is away from the typical line, and one which tends to the loss of everything characteristic of a lizard. In the next forms, the chameleons, the tendency is in exactly the opposite direction, — the specialization of all lizard features.

What more vivid description can there be of the chameleon, the strange form represented upon our plate, than the following of nearly three hundred years ago? "In shape and quantitie it is made like a Lisard, but that it standeth higher and streighter than the Lisards do, upon his legges. The sides, flankes, and bellie, meet togither, as in fishes: it hath likewise sharpe prickles, bearing out upon the backe as they have: snouted it is, for the bignesse not unlike to a swine, with a very long taile thin and pointed at the ende, winding round and, entangled like to vipers: hooked clawes it hath, and goeth slow as doth the tortoise: his bodie and skin is rough and skalie, as the crocodiles: his eyes standing hollow within his head, and those be exceeding greate, one neere unto the other with a verre small portion betweene, of the same colour that the rest of the bodie is: he is alwaies open eyed, and never closeth them: hee looketh about him not by mooving the ball of his eye but by turning the whole bodie thereof: hee gapeth evermore aloft into the aire, and is the onely creature alive that feedeth neither of meat nor drinke, but hath his nourishment of aire onely: about wild fig-trees hee is fell and daungerous, otherwise harmelesse. But his colour naturally is very straunge and wonderfull, for ever and anon he chaungeth it, as well as in his eye, as taile and whole bodie besides: and looke what colour he toucheth next, the same always he resembleth, unless it be red and white."

Still, this picture is not over-accurate; indeed, it makes sad lapses in several respects. A glance at the plate will show the errors in some respects; others will be mentioned. One of the most striking features will be found in the feet. In most lizards the toes are all distinct, but here the foot forms what might be termed a pair of pipe-tongs. The toes are united in bunches, two toes in one bunch, and three in the other. Another peculiarity is seen in the strange eyes covered by the peculiar lids with but a small opening for vision. Stranger still is a feature which the plate does not show. The eyes are entirely independent; that of one side can be turned in any direction without affecting the other. Further it is said, even by good authority, that one side of the animal may be sound asleep while the other is alert; an actual realization of the familiar expression, asleep with one eye open.

The peculiar structure of the feet renders these animals eminently adapted to an arboreal life; for with them they can grasp a limb in the firmest manner. They live, as might be supposed, in trees, wandering about after food. The old author erred when he said that they fed solely upon air. Air is too unsubstantial for them; they demand insects, and are fully equipped for capturing them. As shown in the plate, the tongue is highly extensile, and can be projected to an immense distance from the body. Its tip is very sticky, and to it adheres any insect which may be touched by it. With stealthy motion the chameleon creeps towards a

AFRICAN CHAMELEONS (*Chameleon vulgaris*).

beetle or a fly, and thence suddenly, with the rapidity of light, darts forth his tongue and brings back the insect, the whole operation taking but a fraction of a second.

The chameleons are most renowned for their changes of color, which are much like those of the American chameleon already described, except that they are even more rapid and more extensive. Notwithstanding the old author, they can change to red and white. The change of color in all these forms is brought about by little cells containing different colors. At the will of the animal the color in these can be increased or diminished just as in the squids already described. The change is not wholly dependent on the will; for it is to a certain extent reflex, and will vary according to the surroundings and under such circumstances as to render it probable that it is not always by the volition of the animal. The true chameleons are all confined to the tropical and sub-tropical parts of the Old World, the species figured occurring in the region around the Mediterranean.

TURTLES.

Strange creatures are the turtles, with their bony houses, which they carry around with them. Wherever they are they are at home, and all that is necessary, when they desire seclusion, is to fold away the tail and draw in the head and legs. It is no wonder that they figure extensively in mythology, and what firmer support could have been discovered for the earth than the giant tortoise which the ancients claimed held it up?

This bony investment of the turtle deserves a few words. It is a bony box, usually marked on the outer side into polygonal plates, and showing on the inner surface the backbone, portions of which, together with the ribs, are expanded to form the dorsal part. In its shape and in the character and outlines of the plates great differences can be seen.

First of the turtles is the monstrous marine leather-back, or trunk-back, which is found in all the seas of the world. Although thus widely distributed, it is really a rare species, and the capture of an individual on our shores is chronicled in all of the scientific journals. But little is known of its habits, for naturalists but rarely have a chance to see these creatures, and then generally when they are dead or dying. So far as the writer has been able to ascertain, but fourteen specimens have been taken on our shores since 1824. One of these, a small one, was captured in a mackerel net off Cape Ann in the summer of 1880. For several days the fishermen kept it in a barn, charging a moderate price for admission. It struggled greatly to get back to the water, and its actions were pitiful to see. At last death put an end to its sufferings.

It would appear that this form is rather more abundant in Europe and the east. One author writes that this species is one of those with which the Greeks were well acquainted, and he supposes it to have been the species particularly used in the construction of the ancient harp or lyre, which was at first constructed by attaching strings to the shell of some marine tortoise. This form is very large, having a total length of about seven feet, a spread of flippers of nine or ten feet, and a weight of a thousand or twelve hundred pounds in the largest specimens. Its ridged back

Fig. 354. — Leather-back tortoise (*Sphargis coriacea*).

is covered with a dark leathery integument, and from this appearance the common name is derived. The flesh is very good and palatable, as the writer can testify from his own experience.

The logger-head turtle rarely gets as far north as Massachusetts, but farther south it is very abundant, and at the time of laying the eggs they are taken in large numbers along the sandy shores of the Mexican Gulf. At night they come up from the sea, and excavate shallow holes in the sand, in which they lay their eggs. After this is done they carefully cover up the hole with sand, and now is the turtle-catcher's chance. He rushes down from his hiding-place, and quickly turns the animal on its back, and then proceeds to treat the next one in the same way. When once in this position no escape is possible, for the shell is so broad that no pushing with the feet, or contortions of the neck, will put the creature right side up. The logger-head reaches a weight of nearly five hundred pounds.

The green turtle, that delight of epicures and aldermen, is a much larger species, specimens occurring which weigh eight hundred and fifty pounds. It but rarely occurs north of Cape Hatteras, but south of there, especially in the West Indies and on the northern coasts of South America, it is very abundant. The turtle-catchers capture it in much the same way as they do the logger-head. Besides this, it is often harpooned as it feeds on the sea-weed along the shores. When captured it is not killed, but is kept alive in pens until an opportunity presents itself to send it to market, where, perchance, it may be served as turtle soup, or as likely be used day after day as an advertisement, while the patrons of the restaurant devour that wonderful compound made from the 'mock-turtle,' an animal no naturalist has yet seen and studied in his native condition.

Before our present days of sophistication, when the ways of dyeing horn and the manufacture of celluloid were unknown, the hawk's-bill turtle, or caret, was held in high esteem, for from its armor came all the tortoise-shell of commerce. Says Holland's 'Pliny' (1601): "The first man that invented the cutting of tortoise-shells into thin plates, therwith to seele beds, tables, cupboards, and presses, was Carbillius Pollio, a man verie ingenious and inventive of such toies, serving to riot and superfluous expense." The method of obtaining the plates varies in different localities. In some, the animal is killed, and then the shell is boiled; in others, the plates are started by heat, and are then torn off, the animal being allowed to escape again into the water. Under the influence of heat the plates can be pressed into any desired shape, and they may even be welded together, so that pieces of shell may be obtained far larger than any of the plates occurring on the turtle.

But few queerer-looking animals can be found than the soft-shelled turtle of the United States. Look, for instance, at the peculiar face depicted in our plate, with its oblique eyes, full cheeks, and pointed nose. The species is a small one in comparison with those which have gone before; but none of them equal it in ferocity. A shell twelve inches in length is perhaps the average. The flesh is very palatable, and as the name indicates, the shell is very soft. One of its most marked peculiarities is its power of staying under water. Specimens have been watched steadily for ten hours at a time, and during that period they evinced not the slightest desire to go to the surface to breathe. Their lungs are not large, and so it would seem that they must have the power to extract oxygen from the water, for the supply that they carry down with them could hardly last that length of time. Such is really the case; and Professor Gage has ascertained that in the throat are gills, and he has proved that by means of these gills they actually breathe.

It is a curious feature often noted, that the same discovery is frequently made by different observers, entirely independent of each other, at about

FIG. 355. — Hawk's-bill turtle, or caret (*Eretmochelys imbricata*).

the same time. In every book it is stated that the turtles breathe only by lungs, but here is one which has gills in addition; and as soon as Professor Gage had made his announcement in this country, a naturalist in Australia made the same discovery concerning a turtle in that far-off land.

Rivalling, or even excelling, the green turtle, comes the terrapin of the southern states. Several species figure under this name, and unless the purchaser know what he is after, he is apt to have some other species put upon him instead of the true salt-water species which he desires. It varies considerably in its appearance; above, it may be a greenish gray,

SOFT-SHELLED TORTOISE (*Aspidonectes feroz*).

and between this and a dark brown almost every gradation may be found. The under surface may be either light yellow or reddish brown, plain, or striped, or spotted. The true terrapin lives in the salt marshes, from New York south to Brazil; but is most abundant in the Carolinas and Georgia, where a large proportion of the catch for the market is made.

In the fresh water are many turtles which often figure as terrapin; but to enumerate them would take more space than we can spare, and besides, the description would possess no features of interest. Not all of the group are confined to the water; the wood-tortoise, for instance, being found in the woods and forests, often at points remote from any large body of water.

Among the strangest of turtles are the box-tortoises. There are several species which differ not only in their colors and markings, but also in the

FIG. 356. — Wood-tortoise (*Chelopus insculptus*).

degree in which they are capable of closing the box. With ordinary turtles all that is possible is to draw the head, legs, and tail underneath the shell; but in the box-tortoises, after this is done, the two ends of the ventral plate are capable of being folded up against the upper half of the shell, much like the lids of a box, thus giving a more complete protection.

Our figure shows the most common species of box-tortoise in the eastern United States. It is a rather small species, seldom exceeding seven inches in length of shell. It is eminently a terrestrial species, not liking the water, but often seen wandering slowly along through the meadows, or in the woods, in its search for food, which is wholly vegetable in its nature, consisting largely of toadstools, mushrooms, and similar objects.

While the box-tortoises are inoffensive and depend solely upon their shells for protection, the snapping-turtles are incapable of such defence and have therefore to depend upon their jaws. Their shell is so small that the head and legs cannot be completely concealed as in most common forms, while the long, scaly tail projects far from the body. The strong jaws serve not only for defence, but are often used as organs of offence. While most turtles are of a timid disposition, these are always warlike; and it is said that the moment they emerge from the egg their jaws are viciously snapped, much as in the adults. The old and large specimens

FIG. 557. — Box-tortoise (*Cistudo carolina*).

are very strong. They lurk in ponds, preferring localities along the weedy shores, where they lie in wait for frogs and fishes, their favorite food. When one ventures too near, there is a sudden extension of the neck, a snap of the jaws, and all is over. These jaws are so strong that they can easily cut off a finger of a man, and Agassiz states that he has seen them bite off a piece of a board an inch thick. This species grows to a length of about three feet, and furnishes very good meat.

The alligator-turtle of the south is still larger, but owing to a strong musky flavor the flesh is not good for food. The largest specimen of which I find record measured nine inches between the eyes; its other dimensions were not given, though it was estimated to weigh more than one hundred pounds. Mr. Fontaine thus describes the habits of a specimen which he kept in his fish-pond: "One day, after he had eaten, he remained upon the rock where I had fed him, and which was only about a foot beneath the surface, where it shelved over water ten feet deep. A swarm of minnows and perch were picking up crumbs around him, apparently unconscious of his presence. His head and feet were drawn

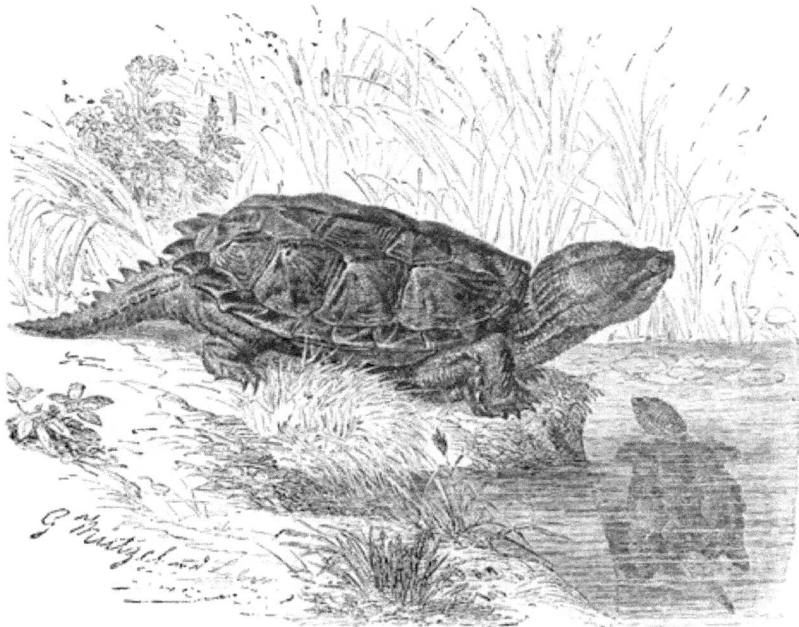

FIG. 358. — Snapping-turtle (*Chelydra serpentina*).

sufficiently within the shell to be concealed. His mossy shell could not well be distinguished from the projections of the rock on which he was lying in ambush. Several large bass were gliding around him, occasionally darting at the minnows. One of these, about fourteen inches in length, came within striking distance of his head, which he suddenly thrust out and fastened upon him, fixing his aquiline bill deeply into his side and belly. He immediately drew the fish under him, and, holding him down firmly to the rock with his fore feet, ate him greedily, very much as a hawk devours his prey."

This musky flavor of the alligator-turtle is not confined to that species. It is even stronger in a much smaller species common in ponds all through the eastern part of our country. Indeed, so strong is the odor that the common name given it is musk-turtle, or stink-pot; and its scientific name, *Aromochelys odorata*, repeats twice classical words for smell.

The name gopher is indiscriminately applied to two very different animals. One of these is a mammal allied to the rats, and will be described and figured later in this volume; the other is a turtle common in the southern states, where it lives in large communities and digs burrows much like those of its namesake of the prairies. Both are regarded as pests. The turtle-gopher is a vegetarian. At night it leaves its burrow and wanders to the plantations, where it commits sad havoc with the plants, especially those of a soft or succulent nature. It will dig beneath the soil to get at the sweet potatoes, while a melon-patch is as attractive to it as it is to a negro. On account of its depredations war is waged against it. The burrows are found, and in front of each a deep pit is sunk. The gopher, on emerging, tumbles into this, and being unable to climb up the steep sides, falls a victim to the negro when he makes his rounds in the morning.

A near relative of the gopher is the gigantic tortoise, which once was very abundant in the Galapagos Islands. It has a large, high-arched shell, and a neck of almost serpentine character. Specimens have been taken with a total length of six feet and two inches, and a weight of six hundred and forty-two pounds. The nearest relative of this giant is to be found in Madagascar. It is almost equally large. Of late years both these species have been greatly reduced in numbers, owing to the persecutions which they have suffered. In former years they abounded upon the Galapagos Islands, sometimes there being hundreds in sight at one time. It is a curious fact that none of them occur on the adjacent shores of South America, a few hundred miles away.

The great river-turtle of the valley of the Amazon is a very important form for the natives, as from it they derive a large proportion of their meat as well as other means of subsistence. These turtles have their regular migrations, going in the wet season to the pools and submerged forests of the interior, and on the return of the dry season, making their way to the main streams. In the pools the Indians capture them, either by nets or by means of arrows, which in their structure are much like the harpoons used by the Eskimo. They have a detachable iron head fastened by a cord, some thirty or forty yards long, to the shaft. The turtle is shot with these arrows, the shaft drops out and floats on the surface, and then the Indian in a canoe picks up the arrow and 'plays' the beast much as our northern fisherman would a salmon, until at last he lands him in the boat. Only the females are used for food; the males, which are smaller and far less abundant, being regarded as very unwholesome.

At the dry season the turtles, in their migrations to the main river, stop on certain sandy islands to lay their eggs, and the operation is watched with considerable interest by the natives. The whole matter of turtle-egging is under government control, and sentinels are appointed to watch the laying, and to see that no one interferes before the proper time, so that all may have an equal chance. The eggs are laid at night, immense numbers of turtles coming at the same time from the water, out on the sand. The leaders dig pits, some three feet deep, and lay some hundred and twenty eggs in them, and then the others deposit theirs on top, until all the pits are full, when the sand is replaced, so that but little signs of the eggs remain. During this egg-laying, which takes some fourteen days, the sentinels are very careful not to alarm the turtles; for on disturbance they are very apt to forsake the spot and seek some more secluded place.

At last the time comes for gathering the crop of eggs. Placards are posted on the doors of the church, announcing the date when the egging will begin. At the appointed time hundreds of natives are gathered on the spot, provided with large copper kettles for trying out the oil, and hundreds of earthen jars for storing it. A tax is levied on each person engaged in the operation to defray the cost of the sentinels, and then, at a given signal, all fall to work simultaneously. When all the eggs are dug up, the mashing begins. They are thrown into a kettle or a canoe, and broken either with sticks or by the feet of the children. Next, water is poured in, and the whole left in the sun for a few hours, when the oil, which has stewed out, is skimmed off and placed in the jars. The annual product of oil is very large, and is estimated at about twenty-four thousand gallons. Now, as it takes some two thousand eggs to make a gallon of oil, the total number of eggs annually destroyed amounts to nearly fifty millions, the product of four hundred thousand turtles.

CROCODILES AND ALLIGATORS.

In the older works the crocodiles and alligators were arranged with the lizards; to-day they are placed in a different place. To be sure, they have a long, scaly body, terminating in a long tail, as do the lizards, but in their internal structure they are so different that the naturalist is warranted in regarding them as the highest of all the existing reptiles.

There are three groups of these forms: the gavials, which are confined to the eastern continent; the crocodiles, which occur on both continents; and the alligators and caymans, which are confined to the New World. All are hated and feared, on account of their ravenous habits, and yet they do an immense amount of good by their labors as scavengers. The tropical rivers in which they dwell bring down large numbers of dead animals, and these, unless removed, would breed pestilence in the lowlands. Now the service of these animals comes in; to them a bit of carrion is far more palatable than any fresh meat; indeed, they are said to frequently bury the animals they kill, and wait until decomposition sets in before devouring them.

Best known of all to us is the alligator, that denizen of the rivers and swamps of the southern states. The traveler upon the rivers of Florida and the bayous of Mississippi sees large numbers of them, basking in the sunshine on the shore, or on some old log, looking so limp and lifeless that one can hardly help thinking they are dead. They are, however, soon alarmed, and they quickly crawl into the water, where they are perfectly at home. In Florida, and in other easily accessible regions, they have now become considerably diminished in numbers, owing to the persistency with which they have been hunted to supply the demand for alligator leather. In the more remote and inaccessible localities they still flourish, and are constantly on the wait for some object of prey. They but rarely attack man voluntarily, unless impelled by hunger, but when aroused they prove no mean antagonists. Their jaws are armed with a fearful array of teeth, while a blow from their strong tail may break a man's leg.

They are said to be especially fond of the flesh of dogs, and to congregate at places where they hear a dog bark, or a puppy whine, with only the tip of the snout and the vicious-looking eyes above the surface of the water. It is said that the natives of the West Indies and South America are well aware of this peculiarity, and when they wish to cross a stream infested with these monsters, they cause a dog to bark at one spot, and when all the alligators are congregated there, they cross, without the slightest danger, at a point above or below. One peculiarity is the fondness they exhibit for swallowing substances from which they cannot possibly extract the slightest nutriment. It is said that this is done to keep the stomach distended, and the fact that sticks, bottles, stones, etc., have been taken from them lends countenance to the view. Our alligator but rarely exceeds twelve feet in length. It lays its eggs in the sand, usually forming a small mound for the purpose.

The caymans belong to South America. There they abound in the rivers. They have their migrations. When the wet season comes, they travel to the pools and flooded forests, which line the banks of the Orinoco and the Amazon; but when the dry season returns, and the waters fall, they descend to the main river. Bates even declares that he has seen the upper Amazon "as well stocked with large alligators, in the dry season, as a ditch is in England with tadpoles." This form builds conical nests of dry leaves, and lays in them about twenty eggs. Waterton's adven-

Fig. 359. — Alligator (*Alligator lucius*).

tures with one of these caymans has often been told, but it will bear one more repetition. He caught one of these animals on the Orinoco, and getting astride of it rode it as one would a horse, until he thoroughly exhausted it.

It was not until 1870 that any crocodiles were known to inhabit the United States, although they were well known in the West Indies and in South America. In that year Dr. Wyman described a skull from Florida,

and since that date several specimens of the species figured here have been
found there. A glance at our figures will show the difference between
the two. In the alligator the snout is broad; in the crocodile it is narrow
and more pointed.

Our crocodile, however, is not so well known as the celebrated inhabitant of the Nile, of which pages might easily be written. There it has

Fig. 360. — American crocodile (*Crocodilus americanus*).

lived from time immemorial, and the ancient Egyptians carved its effigy
upon their monuments, and embalmed its body and placed it in their
tombs. In some places they even regarded it as sacred, and kept it at the
public expense. "It was fed and attended with the most scrupulous care;
geese, fish, and various meats were dressed purposely for it; they ornamented its head with earrings, and its feet with bracelets and necklaces of

gold and artificial stones." In other places it was held in the same detestation as to-day. It now rarely occurs north of Manfalert, but it formerly
descended the Nile, nearly to the sea.

The gavials live in the East Indies, and as they are protected by the
various native religions, they fairly swarm in localities where the Europeans are few, and in such places they really are very dangerous to human
life. The snout of the gavials is even narrower than that of the crocodiles,
but the tip is swollen into a sort of a ball.

We must say a few words about the fossil reptiles, so wonderful are
some of them. In some the most striking features were the enormous
size. The reptiles of the present time are small in comparison; our largest
pythons and alligators shrink before these monsters. America seemed to be
the especial home of these gigantic forms; and in the rocks of our western
territories, Professors Cope and Marsh have exhumed the remains of
almost numberless specimens. Away back in geological time, when the
deposits of chalk were being laid down, these regions were a shallow
inland sea; and in the waters these animals flourished. Some were whale-
like in shape, and had a length of eighty feet, while others had a backbone six feet in diameter. Some swam freely through the water, while
others walked about in water thirty or forty feet in depth, and were still
able to capture animals swimming on the surface.

Others were more like snakes, with very long neck and tail, and the
legs like flippers; while others must have been terrestrial and able to
browse upon the tops of the trees thirty or forty feet in height. Strangest
of all were the bat-like forms, the Pterodactyls, some of which could
spread their wings to a width of twenty feet, while others had the body
terminated with an oar-like tail. About all that we know of any of these
is from their bones; but these give us evidence of animals utterly unlike
anything existing to-day. Restorations are often made of these forms;
but in all these there must of necessity be a certain amount of guess-
work.

BIRDS.

No one needs a definition of a bird, for there is not the slightest chance for mistaking one of these feathered forms for any other animal ; nowhere else in the whole animal kingdom do we find feathers, and, on the other hand, no bird lacks these structures.   Not only the feathers mark off the birds from all other groups ; there are many other features which at once pronounce them distinct from all other forms which live to-day.   Still birds were not always so isolated as they now appear.   The birds of to-day have many points in common with the reptiles ; but in the ages of the past there were forms — true birds — which approximated far more closely to the group which we have just dismissed.   Some of these points of resemblance are too abstruse for consideration in a popular work ; some will appear in the following pages.

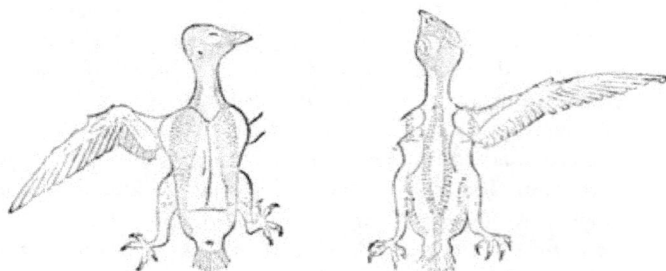

FIG. 361. — Feather-tracts of a swift, showing the regions (dotted) where the larger feathers of the body are inserted.

Since feathers are so characteristic of birds, we may be pardoned a few remarks concerning them, especially since the different kinds of feathers are so much used in classifying birds.   In the first place, feathers are not evenly distributed over the surface of a bird ; there are regions on almost every bird which is without plumage : but the position and extent of these naked spaces (apteria, they are called) varies with the different groups.   In the adjacent cut of the body of a swift, the dots show the points of insertion of the feathers of the body, while on the wings and tail the basis of the feathers themselves are seen.   All the white portions are without feathers, although in life this nakedness does not appear, owing to the fact that the feathers of the other parts spread over them.

In the feather-tracts the feathers arise, each from a small papilla.   The typical feather consists of a shaft or quill, from either side of which the barbs project. In the more familiar forms these barbs adhere to each other by little hooklets, so that they form a firm vane or web. At other times these barbs may be more slender and thread-like, and may remain distinct from each other, thus giving rise to the down seen at the base of almost all feathers, and in some constituting a large proportion

Fig. 362. — A feather from the tail of a king-bird.

or even all of the feather, which is then called a plume. At other times the feathers may consist of a mere thread-like shaft, as in the case of the hairs about the mouth of the night-hawk and similar birds.

So, too, the names of the feathers on the different parts of the body, especially on the wings and tail, have their own importance from a systematic standpoint. In the tail the principal feathers are the tail-feathers, or rectrices; while the feathers which cover their bases above and below are respectively the upper and under tail-coverts. In the wing there are more features to mention. The long feathers which arise from the hand are called the primaries, and those from the fore arm the secondaries, the last three being termed the tertiaries. Sheathing the bases of these long feathers are the upper and under wing-coverts, which are divided into several groups, as shown in the cut, to which reference should be made for details.

It is to the feathers that the birds owe their beauty. No artist's brush can reproduce some of the brilliant hues. In many cases the feathers have a color of their own; but in others, and this is true of some of the most brilliant, the effect is not due to any pigment; but instead is of exactly the

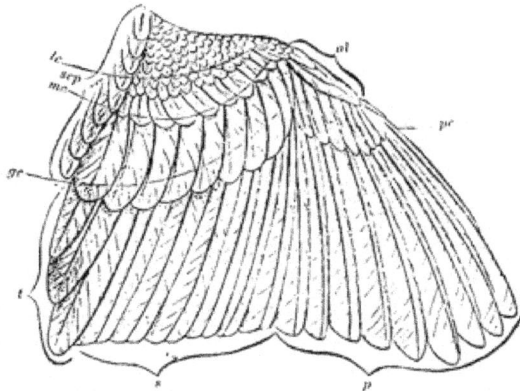

Fig. 363. — Wing of a sparrow: *lc,* lesser coverts; *mc,* middle coverts; *scp,* scapulars; *gc,* greater coverts; *t,* tertiaries; *s,* secondaries; *p,* primaries; *pc,* primary coverts; *al,* ala spuria.

same nature as that described on a preceding page as occurring in mother-of-pearl (p. 79). When the birds hatch from the egg, they may either be entirely naked, or they may be covered with down, or, again, they may have true feathers. Some, when hatched, are abundantly able to take care of themselves; others demand the mother's attentions. At various times during life the feathers are changed. When fully formed, a feather is no longer capable of any alteration, or even of being repaired in the case of any injury; and so this feather-changing — molting, it is called — plays an important part. In some birds the operation is a gradual one, one feather after another being dropped, and new ones taking their place in the same order; while in others, so many may be lost at once as to make the bird unable to fly. With these molts are frequently connected changes in the colors of the plumage, often of great extent.

Of all groups of the animal kingdom there is none which excites the same amount of interest that the birds do. Their graceful shapes, their bright colors, and their interesting habits, render them fit subjects for both the painter and the poet; and the amount of extravagant writing they have inspired is almost incredible. Yet as one studies the resplendent colors of the humming-birds, or watches the majestic flight of the birds of prey, he can pardon any use of poetic license; for, be as extravagant as one can, no description and no figure can equal nature.

This interest in birds has resulted in a literature perfectly enormous, the extent of which one cannot begin to realize until he stands in the alcoves of a scientific library, and sees shelf after shelf and case after case filled with books, some small, others large quartos and folios, all devoted to the description of the appearance and habits of our feathered friends. Some are extremely technical in character, giving long-winded descriptions of feathers and bills, bones and muscles; while others — far more interesting — go to the other extreme, and give the songs, the nesting habits, and long and charming accounts of the psychological side of bird life. These works, too, have an artistic side; for most of them are illustrated with plates, the production of which calls for the highest skill of the artist and the engraver. No other class of books in the whole range of literature equals the books on birds in this respect.

SOME OLD-FASHIONED BIRDS.

One of the rarest and at the same time most interesting of fossils is the curious form represented in our figure. It is rather mixed up; but still it reveals some very interesting features, the most striking of which is the long-feathered tail which extends from the centre of the cut down towards the bottom of the picture. Now, in all modern birds the tail is extremely short, and consists of but a few bones welded together; but here is a bird with a long tail like that of a reptile, except that it is feathered on either side. No living bird has teeth; but in other specimens of this old-fashioned form teeth do occur, and are planted in the jaws, almost exactly as in some of the reptiles. Here, then, is one of the forms which may be said to partially close the gap between the existing reptiles and birds, and yet this form is distinctly a bird. The only specimens known have been found in the large quarries of lithographic stone at Solenhofen, in Bavaria. These are the oldest known remains of birds, the rocks in which they occur being of Jurassic age. The specimen figured was about as large as a crow; another has been found considerably larger.

Our own country also has its fossil birds; forms nearly as strange as the one just mentioned, but of which we know far more than we do of that, owing to the excellent preservation of the remains. These, too, were birds with teeth, but they lacked the long tail of their older European relative. Of these birds two very distinct forms

Fig. 364. — The oldest known bird (*Archæopteryx*). The peculiar tail is most noticeable in this specimen.

are known. One had the teeth implanted in sockets in the jaws, and in this and in many respects was so like a reptile, that if one did not have other parts of the skeleton, he would be as likely to class it among that group as among the birds. This form is known as *Ichthyornis* (the fish-bird); it was about the size of a pigeon, and had a tail much like that of modern birds. Its strong wings would indicate that in its habits it was much like the terns and gulls of the present day.

The other form, *Hesperornis* (the bird of the west) was even more strange, for it was actually a bird without wings; and Professor Marsh even has his doubts whether it could walk upon the land. On the contrary, he regards it as having been far more aquatic in its habits than even the penguins of to-day. It was well adapted for a life of swimming and diving. Its legs and feet formed admirable paddles, and the bones show that it must have had strong muscles to move them; while the bones of the tail, flattened from above, would indicate that that organ was also of use in swimming, and must have been shaped much like that of a beaver. When it lived, the place now occupied by the great plains was a shallow, tropical sea, five hundred miles wide, and no one yet knows how long. It swarmed with fishes, and on these forms, loon-like, the *Hesperornis* doubtless fed. What a strange creature it must have been, with its long snake-like neck terminating in a head and jaws like those of a lizard!

Fig. 365. — A bird with teeth (*Hesperornis*).

OSTRICHES.

Strange creatures are the ostriches and their allies. Indeed, they are to be regarded as almost as 'old-fashioned' as the forms just mentioned. Almost all of their relatives lived in the past, and the few remaining forms seem greatly out of place among the birds of to-day. Yet still they linger in the southern hemisphere, and from present appearances it will probably be a long time before they all become extinct.

First and foremost come the true ostriches of Africa, the largest of living birds. Of these there are probably three or four species, including a doubtful form which occurs far to the north of its relatives in the Syrian

Fig. 366. — African ostrich (*Struthio camelus*).

desert. All are much alike in their appearance and habits, and our figure, for all practical purposes, will answer for all. In reality it represents the best-known species, which is distributed over almost the whole of the

African continent. Its head and neck are naked, or only covered by a sparse hair-like down. The body is feathered — black in the male, brown in the female; while the legs are naked, and the feet have but two toes. The wings are very small and, like the tail, are covered with white feathers — the highly prized plumes of commerce. The adult ostrich stands some seven feet high, and weighs from a hundred and fifty to two hundred pounds.

The first reference to the ostrich occurs in the book of Job. "Gavest thou . . . wings and feathers unto the ostrich? which leaveth her eggs in the earth, and warmeth them in the dust and forgetteth that the foot may crush them, or that the wild beast may break them. . . ." The ostrich is represented on the ancient temples of Thebes, and the Pharaohs had their fans of ostrich plumes. In ancient Rome, two thousand years ago, live ostriches were exhibited to the populace, and the dames of that luxurious city prized the feathers as highly as do the ladies of fashion to-day. And yet, until 1862, there was no serious attempt to domesticate this valuable bird. All the feathers were taken from the wild birds, and no attempts· were made to keep it from extermination. To-day all this is changed; but before considering a modern ostrich-farm let us look a moment at the bird in its wild state.

So many, and so conflicting, are the statements that it is a considerable task to say exactly what the characteristics of the ostrich are. As Mr. Biggan expresses it, "the ostrich is a wonderful paradox. . . . He is capable of scanning the whole horizon, and yet, falling easily into some hole under his feet; he is both a gourmand and an epicure; he may be kept in bounds by a fence of a single wire, yet when panic-stricken, he will risk a collision with a stone wall, and dash himself to death; he is both blood-thirsty and gentle; bolder than a lion, and more timid than a spring-bok; a polygamist and a celibate; capable of extreme parental tenderness, yet sometimes eating his own offspring; at once the stupidest and the most cunning of birds." From these and other conflicting characteristics, together with an enormous amount of exaggeration and imagination, the ostrich of the books has been evolved. To most persons an ostrich is a bird which, when its head is buried in the sand, believes its whole body is concealed; a bird which lays its eggs in the sand, and then leaves them to their fate. In reality, the ostrich possesses neither of these characteristics.

The eggs, about fourteen in number, are laid, either in some natural depression, or in a hollow, which both birds of a pair scrape in the soil, and around which they dig a trench, to protect the nest from the rain. The cock and the hen then alternately sit upon the eggs, — the cock during night — for forty-two days, and then the callow brood is hatched, the operation taking just twice the time that it does with our familiar barn-yard fowl. When they escape from the eggs the chicks are covered with a light-colored down, barred along the back with black. They, like old birds, feed upon grass, leaves, insects, and occasional reptiles; grow rapidly, and assume the adult plumage in the third year. The young ostrich is peculiar in its fondness for dancing. It will run a few steps, then turn around, sometimes five or six times; a few steps more, and then another whirl, and so on, for an hour at a time. An ostrich-hunt is a serious affair, as the following account, by Canon Tristram, will show. The scene is laid in northern Africa.

" The capture of the ostrich is the greatest feat of hunting to which the Arab sportsman aspires, and in richness of booty it ranks next to the plunder of a caravan. But such prizes are not to be obtained without cost and toil, and it is generally estimated that the capture of an ostrich or two must be at the sacrifice of the lives of two horses. So wary is the bird, and so open are the vast plains over which it roams, that no ambuscades or artifices can be employed, and the vulgar recourse of dogged persever-ance is the only mode of pursuit. The horses to be employed undergo a long and painful training, abstinence from water and a diet of dry dates being considered the best means for strengthening their wind. The hunters set forth with small skins of water, strapped under their horses' bellies, and a scanty allowance of food for four or five days, distributed judiciously about their saddles. The ostrich generally lives in companies of from four to six individuals, which do not appear to be in the habit, under ordinary circumstances, of wandering more than twenty or thirty miles from their headquarters. When descried, two or three of the hunters follow the herd at a gentle gallop, endeavoring only to keep the birds in sight, without alarming them, or driving them at full speed, when they would soon be lost to view. The rest of the pursuers leisurely proceed in a direction at right angles to the course which the ostriches have taken, knowing by experience their habit of running in a circle. Posted on the best lookout they can find, they await for hours the anticipated route of the game, cal-culating upon intersecting their path. If fortunate enough to detect them, the relay sets upon the now fatigued flock, and frequently succeeds in running one or two down, though a horse or two generally falls exhausted in the pursuit. The ostrich, when overtaken, offers no resistance besides kicking out sideways."

In 1862 the first attempt at ostrich-farming was made in the Cape Colony. Three years later only eighty tame birds were reported; but the success was so great that in 1875 it was estimated that there were fifty

thousand ostriches in confinement in South Africa; and since that date
the number has more than doubled. In 1875 the feathers reported were
worth about $2,000,000; and in the years 1881 to 1883 the value of the
'crop' was estimated at about $5,000,000 a year from South Africa alone.
In addition, Egypt exports feathers to the amount of about $1,250,000;
and the Barbary States only about $100,000. Besides, there are now
ostrich-farms in Buenos Ayres and in southern California, which give
promise of being very successful. The following, the latest statistics at
hand, apply to the year 1880–81. A hen will lay about ninety eggs in a
year; and from these about sixty chicks will hatch, worth on the emergence from the shell about twenty-five dollars apiece; and at six months
their value increases to about seventy-five to a hundred dollars a bird.
Since that time the value of the ostriches has greatly declined; and a pair
of breeding birds in 1884 were worth on the average only about one
hundred dollars.

On an ostrich-farm three or four broods are raised a year. The birds
are fed on grain and on the fleshy leaves of the prickly pear. In preparing
the latter the prickles are singed, and then the leaves are cut into pieces
an inch or two square. How long an ostrich lives is as yet an uncertain
quantity. Some place the average life at twenty-five, others at fifty years.
When the time for picking the feathers comes round, a bag or stocking is
drawn over the head of the bird; and the plumes are cut off about two
inches above the root of the quill; and about three or four months later
the dried stumps are pulled out with a pair of pincers. In the wild state
the feathers are renewed but once a year; but in confinement by means of
this cutting and plucking out the roots three pluckings in two years are
obtained. At first the feathers were pulled out; but this was found to be
very detrimental, as the next feathers were almost invariably distorted.
After picking, the feathers are assorted into different grades, according to
color and condition, and then bailed up; they are sent to Port Elizabeth
or Cape Town, the former being the largest market for feathers.

In the wild state, even in the breeding season, the ostrich but rarely
attacks man. But on every ostrich-farm there is usually at least one
vicious bird for whom the farmer must always keep on the watch. Says
Mr. Biggan, in a most interesting article on the ostrich: "When one
approaches a vicious bird's camp at the breeding season, the cock exalts
his head and body and, coming toward the stranger with stately and very
deliberate strides, begins to hiss loudly like a goose or serpent, at the same
time erecting all his feathers and spreading his wings till he becomes twice
his usual size. When perhaps twenty yards off he drops suddenly on
his knees, appearing, as it were, in a sitting posture. Curving his neck

haughtily over his body, he swings it swaggeringly from side to side, at each movement knocking his head violently against his body. In this performance he partly fills his throat with air, so that every thud is accompanied by a peculiar gurgling sound; while keeping time to these movements, his great wings swing alternately backward and forward in a boastful manner. This is called the 'challenge.' It is well named, for there is a bragging, tread-on-the-tail-o'-me-coat air about it that would be irresistibly laughable — if only it could be seen from the safe side of a tall fence, instead of over the low barrier of dried bushes, of which most camps are composed. . . . If, however, the bird gets near enough to his opponent to give the so-called kick, he lifts his bony leg as high as his body, and brings it down with terrible force. His object is to rip the enemy down with his dangerous claw; but in most cases it is the flat bottom of his foot which strikes; and the kick is dangerous as much from its sheer power as from its lacerating effects. It is a movement of terrible velocity and power, at all events. Several instances may be mentioned of herd-boys being thus wounded, maimed, or killed outright. One case occurred near Graaff Reinet, in which a horse had its back broken by a single blow. In this case the bird had endeavored to kill the rider, but missed him and struck the horse." When attacked, and without a tree or other shelter, the keepers usually lie flat on the ground, so that the bird cannot strike them. But even then there is trouble, for the bird will trample on him and roll over him in the most contemptuous manner. The drivers usually arm themselves with a thorny stick, which they present to the neck, the most sensitive portion of the enraged bird, and thus easily keep him out of striking distance.

The ostriches of South America differ considerably from those of Africa, but the only peculiarity which we need to mention is the possession of three toes to the feet. There are three species of them, but all are grouped under the common names of nandu and avestruz. In their distribution they range from southern Brazil to the Strait of Magellan. In height, they stand from five to six feet; in color, the common form is brownish gray above, and nearly white upon the belly. Their feathers are worth but little in comparison with those of their African cousin, and still a bird may produce five or six dollars' worth when sold in the home market. They are strong runners, but do not equal the true ostriches in this respect, and hence are more easily hunted.

In Australia the emu represents the ostrich group, and, like its relatives already mentioned, it lives on the plains. Formerly it was very abundant; but now, owing to a merciless persecution, it has been driven from all the settled portions of that island-continent. It is hunted largely for 'sport';

for the flesh is not over-palatable, and only the hind-quarters are ever eaten. The cassowaries, on the other hand, are inhabitants of the forests. They, too, are large birds, one occurring in Australia, the other nine in New Guinea, Ceram, and the adjacent islands. The most striking feature in these birds is the helmet or crest on the top of the head, which does not attain its full development until the fourth or fifth year. In both emus and cassowaries the wings are much more rudimentary than they are in the true ostriches.

A few extinct birds demand a moment's attention. First comes the moa of New Zealand, which must have been exterminated since that island was peopled by the Polynesian Maori. Indeed, there is some evidence to show that individuals were alive as late as the last half of the last century. A few feathers, a few bits of dried skin and muscles, are all that we now know of these forms, except the immense numbers of bones that exist. From these remains some dozen or fifteen species have been indicated ; the largest, with truly elephantine legs, reaching up to a height of twelve or fourteen feet. Their eggs, too, have been found, — enormous objects, dark green in color, oval in shape, with a length of ten inches and a breadth of seven.

About 1850 some natives came from Madagascar to Mauritius to buy rum, and wonderful were the receptacles they brought with them to contain the liquor. They were nothing but the egg-shells of some unknown bird of enormous size; for each egg would hold about two gallons. A search was made, and remains of three species have been brought to light. *Æpiornis* is the name which has been given to them, but the size of the largest did not much exceed that of the ostrich, although the egg was six times the size of that of the African bird. In all probability these forms are extinct. It may be that the last was killed some two hundred years ago.

Nearly as strange as any of the forms mentioned are the peculiar wingless kiwis of New Zealand, small nocturnal birds (the largest no larger than a turkey) which feed upon the gigantic earthworms of that country. They are dull brown or gray colored forms with long, pointed beaks. Their eggs are enormous in proportion to the size of the bird, and will weigh about a quarter as much as the bird itself.

One more group of forms, and we are done with the allies of the ostriches. These forms are the tinamous and similar birds, the fifty species of which range from Central America south to Patagonia. The largest species, occurring in Buenos Ayres, is a little larger than a partridge, and is so grouse-like in its appearance and habits that it has won for itself the common name perdiz grande, large partridge. It lives in

CASSOWARY (*Casuarius casuarius*).

the thick grass of the savannahs, and is but a feeble flier, making, when it rises, a whirring noise similar to but much louder than that of our own grouse.

PENGUINS.

The penguins and rock-hoppers have been aptly termed the seals among the birds; for they are almost as aquatic as the seals, while their wings, covered with scale-like feathers, have been modified into flippers, of great use in swimming, but utterly useless as organs of flight. All dwell in the southern part of the southern hemisphere. Our plate represents the best-known form, the king-penguin; but most of the specimens are figured in a position they but rarely assume on land. The head and neck should be stretched straight upwards, the tip of the beak pointed towards the zenith. When on the shore, they are the perfect picture of awkwardness, but in the water all is changed. They swim with the utmost readiness, and go long distances beneath the surface, rising now for air, and then diving again. Indeed, their motion in the water may be truly called an aquatic flight, for the wings are even more important than the webbed feet. During the day they are almost always in the water, except when sitting on their eggs, but at night they retire to the shores. In habits all the penguins and rock-hoppers are much alike, and the following account of a rock-hoppers' rookery will answer almost equally well for all. It is by Professor Moseley of the 'Challenger' expedition.

"It is impossible to conceive the discomfort of making one's way through a big rookery. You plunge into one of the lanes in the tall grass, which at once shuts the surroundings from your view. You tread on a slimy, black, damp soil, composed of the birds' dung. The stench is over-powering, the yelling of the birds perfectly terrifying. The nests are placed so thickly that you cannot help treading on eggs and young birds at almost every step. A parent bird sits on each nest, with its sharp beak erect and open, ready to bite, yelling savagely *caa, caa, urr, urr*, its red eye gleaming, and its plumes at half-cock, and quivering with rage. No sooner are your legs within reach than they are furiously bitten, often by two or three birds at once. . . . At first you try to avoid the nests, but soon find that course impossible; then, maddened almost by the pain, stench, and noise, you have recourse to brutality, and the path behind you is strewed with the dead and dying and bleeding. But you make miserably slow progress, and, worried to death, at last resort to stampeding as far as your breath will carry you." The penguins feed on the fish, crabs, and molluscs which abound in the shores of the Antarctic seas. Both sexes aid in incubating the eggs.

KING-PENGUINS.

SWIMMING BIRDS.

Recent studies of the anatomy and development of the 'water-birds' have sadly upset the old ideas of classification, and the result is a complete re-arrangement of these forms. But since this new system is very technical, the old grouping may still be retained, and all the water-birds, except the penguins, divided into two large groups, — the 'swimmers' and the 'waders,' — by taking into consideration the most marked peculiarities in their habits. The details of the new classification should be sought in technical works.

First in order come the grebes, strange-looking and strangely constructed birds, which are more aquatic in their habits than any other birds, except the penguins. Their plumage is soft and silky below, and the head, in the male of most species, is curiously ornamented, during the breeding season, with crests of brightly colored feathers, giving them a most bizarre appearance. They breed solely in fresh water, and build floating nests out of sticks and leaves, fastening it to the rushes on the margins of the stream or pond. There are several species in the United States.

FIG. 267. — Loon (*Urinator imber*).

Much better known is the loon, or great northern diver, whose harsh scream wakes up the echoes of our northern ponds and lakes. It is a beautiful bird, with its dark green and black plumage relieved with streaks and spots of white, the white lines upon the neck standing out from the surrounding feathers in a most pleasing manner. A diver it truly is, and every sportsman tells of the difficulty he experiences in getting a shot at it. No sooner does the gun flash than the loon is beneath the surface, and then when he reappears it may be he is a quarter of a mile away. Usually but a pair of birds occupy a pond, and only for the breeding season. Their nest is made in the damp earth, on the sedgy margin, and in it are deposited their dark greenish brown eggs.

The auks are far more numerous, and all the forms are confined to the northern hemisphere, most of them being restricted to its colder portions. In their appearance they vary much, but all are clumsy-looking creatures.

Of the true auks, the great auk is, without doubt, the most celebrated. We have already seen that some birds have been exterminated by the hand of man, and in the subsequent pages others will be mentioned, and yet one in thinking of extinct birds invariably brings up first the large, clumsy

Fig. 368. — Great auk (*Plautus impennis*).

bird of the northern Atlantic shores. It was a large form, standing thirty inches high, but possessed no especial beauty, its colors being a brownish black and white. The whole interest surrounding it centres in its fate. A hundred years ago these birds were abundant; to-day, in all probability, not a single individual is alive. It formerly extended south as far as Massachusetts, and afforded the Indians many a good meal, as the shell-heaps testify. Its last appearances were at the Funk Islands (just north of New-foundland) and in Iceland. In the former locality, about 1825, it was very abundant, but soon became exterminated, as it was hunted for its feathers. The fishermen would surround a flock of these flightless birds, drive them on shore into stone pounds, and then kill them. To remove the feathers, they immersed the birds in scalding water, and so fat were the bodies that they were burnt to heat the kettles. The Iceland colony was exterminated in 1844.

To-day relics of the great auk are highly prized, and a recent inventory (1884) enumerates, in all the museums of the world, nine complete or

FIG. 369. — Group of least auks (*Simorhynchus pusillus*).

nearly complete skeletons, sixty-eight eggs, and seventy-six skins or mounted birds. Of these, five birds are in American museums. The American Museum in New York was the last to receive a specimen. It cost the donor six hundred and twenty-five dollars.

The rest of the family of auks is made up of numerous species and innumerable individuals of auks, sea-doves, guillemots, murres, puffins, and the like, only a few of which can be mentioned. In the northern seas they occur in enormous numbers, some having a very wide range, while others are restricted to a more limited district. Of these latter, the least auks, represented in our cut, inhabit the region around Bering Strait. They are the smallest of water-birds, not larger than a common robin. They are black above, white below, while on the top of the bill is a curious knot, which appears each year before the breeding season, and then is shed when that period is over. A little larger is the crested auk from the same seas, with its curious bunches of feathers projecting from the sides of the head, and the "little curl right in the middle of the forehead." Its plumage is brownish black above, ashy beneath, and the short little beak is scarlet or orange. All of these forms resort to the shore during the breeding season, but as soon as that is over, they retire to the open ocean, some of them going to warmer climates to spend the winter.

Fig. 370. — Crested auk (*Simorhynchus cristatellus*).

The murres or guillemots, of which there are several species, must also be mentioned. They, too, have similar habits, spending the winter on the open ocean, and coming to the rocky shores in the early spring to lay their particolored eggs. They are especially fond of nesting on the ledges of the cliffs, which are hardly wide enough to hold them. Here they sit in long rows, each bird with its face to the cliff as it towers above them, their black backs turned to the open sea. Says Dr. Stejneger: "When flying off the nest they consequently are compelled to first turn round, and, if taken by surprise, this manœuvre will often cause them to throw the egg from the shelf into the water. It happened several times, when I stealthily approached in a boat under the breeding colonies, that several eggs were thrown into the boat, when the birds rushed off the nest, if the bare rock on which the egg is placed can be called a nest, and my Aleutian oarsmen were always in a roar of laughter when one of these projectiles exploded on the head of an unfortunate companion." At other times the murres seek a more level space for raising their young, and then the shore presents a singular sight. Everywhere can be seen the birds sitting bolt upright upon their eggs, while thousands more are circling around in the air. The eggs are highly prized by the natives, and in the breeding season they form a not inconsiderable part in the diet of the inhabitants of northern climes.

The puffins are also northern birds whose strangest feature is seen in the curiously ridged and even distorted beaks. Best known is the common puffin or sea-parrot of the Atlantic, a bird about a foot in length, brownish black above and dirty white below. Its yellow bill is crossed by three or

four curved grooves.   Stranger still are the horned puffins and tufted
puffins of the Pacific shores, the latter descending as far south as the
Farralones. those rocks which guard the entrance to San Francisco Bay.
Farther north. among the Aleutian Islands. they occur in immense numbers. and are eagerly hunted by the natives.   Their flesh and eggs are
used for food. while the skins. turned feather side in. are made into cloaks
or other garments.   The birds are. like most water-birds, very fat ; and to
remove the oil from the skin a peculiar process is resorted to : " they are

Fig. 371. — Gathering murres' eggs in Alaska.

chewed over and over again by the women and children until all the fatty
matter has been chewed out, that being their method of tanning."
     About the first of May the tufted puffin makes its appearance on the
islands of the Bering Sea in incredible numbers. and then is the opportunity of the Aleuts.   They fly fast but low ; and a party of bird-catchers
makes one think of a number of crazy entomologists.   Each is armed with
a large net. stretched on a hoop four feet in diameter, and supported on
poles ten or twelve feet long.   As the birds fly past. this nest is frantically
waved in the air just as the insect-hunter does when a butterfly passes
him ; and each stroke is apt to bag one of these birds.
     The gulls and the terns are closely allied ; but in most cases they may
be separated by the character of the tail, — forked in the terns, even or

wedge-shaped in the gulls. Still this distinction does not always hold good. Gulls are found in all parts of the world, especially in the neighborhood of large bodies of water.
Some are moderate in size, while others are among the largest of aquatic birds. All along our shores these beautiful birds with their graceful flight form an important element in the landscape, some circling round and round high in the air, others walking along the beach in search for food. In their colors there is not much variety. White predominates, while across the back and wings, like a cloak or mantle, is a patch of darker color, varying from a light gray through all the intervening shades until an almost glossy black is reached. The species are numerous; but all are much alike in habits, — loud-voiced, quarrelsome crea-

FIG. 372. — Horned puffin (*Fratercula corniculata*), on the left; tufted puffin (*Lunda cirrhata*), on the right.

tures, ready to rob any more industrious bird of its hard-earned dinner.

Of all the gulls none is more interesting than that figured, — Ross's gull, an inhabitant of the far north; and this interest centres not in its habits, but in its great rarity. Until 1879 there was not a single specimen of this form in any American collection, while in Europe there were less than ten. In that year, however, several were obtained for the National Museum; and three years later a large number were shot at Point Barrow. Of the specimens taken in 1879 a note is demanded. The naturalist of the 'Jeannette' expedition shot eight of these birds north of Siberia. At last the 'Jeannette' was lost, and the crew had to take to the boats. Everything that they took with them "had to be weighed literally by the ounce"; but still Mr. R. L. Newcomb, the naturalist, clung tightly to three of his skins, brought them safely to the Lena delta, carried them across Siberia, and at last placed them in the Smithsonian Institution.

The terns, or sea-swallows, are much like the gulls in their habits, and yet they exhibit peculiarities of their own. They are almost constantly on the wing, turning so gracefully, and flying so easily, that one cannot restrain his expressions of delight and admiration as he watches them. Now one is sailing slowly and elegantly through the air, when he espies a school of fishes in the water below. Instantly all is changed. The circling flight is stopped, and like an arrow the bird plunges downward to, and even beneath, the waves, and when he mounts upwards again, giving a flirt, to shake off the water, he almost always has a fish.

FIG. 373. — Ross's gull (*Rhodostethia rosea*).

There are numbers of terns belonging to the American fauna; some large, some small, but all loud-voiced, and all curious. Some may be found near the sandbars of our southern rivers, while others occur in the valley of the Mississippi, and still others range north into the Arctic seas. Our largest species, the Caspian tern, is shown in the cut, and in a general way represents the appearance of all the group.

The terns of the Arctic seas have their enemies in the jægers, sea-hawks, or, as the sailors call them, the 'teasers.' These are large birds, and of one of them Macgillivray, an English naturalist, wrote, many years ago, a most interesting account. "The sea-birds are on the wing, wheeling and hovering all around, vociferous in their enjoyment, their screams mingling into one harsh noise, not less pleasing, for a time, than the song of

the lark or blackbird. Every now and then a tern dips into the water,
and emerges with a little fish in its bill, which it swallows without alighting. In the midst of all this bustle and merriment, there comes gliding
from afar. with swift and steady motion. a dark and resolute-looking bird.
which, as it clears a path for itself among the white terns, seems a mes-
senger of death. But a moment ago he was but a dim speck on the horizon. or at least some miles away, and now, unthought of. he is in the very
midst of them. Nay, he has singled out his victim, and is pursuing it.

FIG. 374. — Caspian tern (*Sterna tschegrava*).

The latter, light and agile, attempts to evade the aggressor. It mounts.
descends, sweeps aside, glides off in a curve. turns, doubles. and shoots
away, screaming incessantly all the while. The sea-hawk follows the
frightened bird in all its motions, which its superior agility enables it to
do with apparent ease. At length the tern, finding escape hopeless. and
perhaps terrified by the imminence of its danger. disgorges part of the
contents of its gullet, probably with the view of lightening itself. The
pursuer, with all his seeming ferocity, had no designs upon the life of the

poor tern; and now his object is evident, for he plunges after the falling fish, catches it in its descent, and presently flies off to attack another bird." The teaser is not compelled to act this part of a pirate; for he is a strong swimmer, and apparently as well adapted to fish as is a tern or gull. Neither can it be economy of labor to thus live by highway robbery; for, as our author remarks, the trouble of compelling the other birds to disgorge is apparently greater than that which would abundantly supply it with an honest livelihood.

Among the strangest of birds' beaks must be mentioned that belonging to the skimmers, or shearwaters, long-winged birds, of which one species occurs on our coast. The bill is like a pair of shears, the upper blade being stronger and shorter than the knife-like lower one. As the bird flies over the surface, in a most curious and erratic manner, turning hither and thither with the utmost

FIG. 375. — Bill of the black skimmer.

ease, this lower blade of the scissors ploughs the water, and occasionally scoops up some small fish, which are quickly held by the closing of the upper jaw.

Largest of water-birds are the albatrosses, those strong-winged creatures, which day after day follow the ships, as they sail over the tropical oceans, and the seas of the southern hemisphere. No bird exceeds them in powers of flight. They seem to require no rest, and as one watches them circling round and round, hardly the slightest motion of the wings can be seen. It almost seems as if they are as light as the air in which they live. Now they are close behind the ship, stooping to pick up some offal, thrown from the cook's galley; a few moments later, and their circling flight has carried them miles away. They have not lost the ship, however; for soon they are back again, flying, or rather sailing, in the same tireless manner. No matter how fast the ship may sail, they have not the slightest difficulty in keeping up with it, or even in flying round and round it, in circles miles in diameter. Night comes on, and these birds are lost to sight, but in the morning they are seen again. Do they never rest? Do they follow the ship all night on tireless pinions?

Nobody knows with certainty. Still the probabilities are strong in favor of the following view. In the day-time, the numbers of these birds following a ship is very large, sometimes mounting up into the hundreds. At night, looking over the stern, a few can frequently be seen; but nothing like the numbers visible during the day. So, too, in the early morning, but few are visible; but the numbers rapidly increase as the day wears on. Birds have been caught, daubed with paint, and then set loose, to be recognized following the ship for several days. The probable view is that the birds, after dark, settle down on the water to rest, and with the morning's dawn they mount to a height of some hundreds of feet. "A height of one thousand feet would enable a bird to see a ship two hundred feet high, more than fifty miles off; and often, although unable to see a ship itself, it would see another bird which had evidently discovered one, and would follow it in the same way that vultures are known to follow each other."

FIG. 376. — Sooty albatross (*Diomedea fuliginosa*).

In this way it might regain the same ship it followed the day before, or again, it might get sight of another one, and so lose the first entirely.

There are several species of albatrosses. The largest is the wandering albatross, the wings of which sometimes spread fourteen feet. Much smaller are our northern species. All, however, afford much sport to the traveler, tired by the monotony of a long voyage. A hook baited with pork is thrown over the stern into the wake of the ship, and soon an albatross is caught and hauled to the deck, where every particle of its native grace disappears, and the bird appears the perfect picture of awkwardness, as it is utterly unfitted for locomotion on a hard, flat surface like a ship's deck.

Near the albatrosses comes the long series of petrels, which vary in size from the small 'Mother Carey's chickens,' scarcely six inches long, to the giant fulmar three feet in length. All of them live on the high seas, and many of them will follow a ship just as do the albatrosses. They, too, are regarded with a superstitious reverence by the sailors, and sure disaster must follow any injury to any of these birds. The little 'Mother Carey's chickens' are especially the friends of the sailor;

and one never tires of watching them as they skim along just over the surface of the sea, now rising to the crest of the wave, and then descending into the trough.

With the tropic-birds we take up another series of swimming birds, with very different features from those already mentioned, and which have a structure which points in the direction of herons. The tropic-birds them-

FIG. 577. — Red-billed tropic-bird (*Phaëthon æthereus*).

selves are well named, for they are chiefly found in the seas of the tropical regions, sometimes hundreds of miles from land. They have large wings; but their flight is labored, the strokes being rapid and incessant, and the long middle tail-feathers — red in some species, white in the others — trailing behind.

Allied to these are the frigate-birds of the same regions, of which only two species are known. They are also called man-of-war hawks and hurricane birds, the latter being expressive of the rapidity of their flight, which is scarcely excelled by that of any other bird. Every one that sees them writes the most enthusiastic description of their habits; and we cannot forbear quoting the description given by Mr. H. O. Forbes, as he saw them in the Cocos Keeling Islands of the Indian Ocean. He was looking at the noddies and gannets, and "watching what has been described over and over, but was new to me, how their industrious habits are taken advantage of by the swift-winged frigate-birds. Hiding in the lee of the cocoanut-trees, the frigate-birds would sally out on the successful fishers returning in the evening, and perpetrate a vigorous assault on them, till they disgorged for their behoof at least a share of their supper, which they caught in mid-air as it fell. Such feelings of reprobation as I ought to have felt at their conduct was, I fear, not very deep; for the swoop after the falling spoil was so elegant an evolution, that I confess I always hoped that the poor noddy would give up as heavy a morsel as possible in order to necessitate a correspondingly eager dive after it. Refractory gannets were often seized by the tail by the frigate-birds, and treated to a shake that rarely failed of successful results. Fierce foes as they were in the air, on *terra firma* they roosted near each other like the best of friends. The islanders tame the frigate-birds, and use them as decoys. A hunter wishing to shoot a few of these birds, throws out within gunshot on the surface of the water a piece of attractive bait, which the tame frigate-bird swoops down upon, almost ostentatiously, time after time, to pick up. Several of its hungry brethren, always hanging about, soon make their appearance to struggle after a share; after two or three gyrations. the eager stranger swoops down for the tempting morsel, the decoy soars out of reach, while the unfortunate dupe falls a victim. If the others take flight, the same tactics will be followed again and again by the decoy, who exhibits no alarm at the report of the gun, or the death-throes of its companions." Not only do they rob the noddies and gannets; they even steal the fish from the fish-hawks in the Gulf of Mexico.

Frequently the stolen booty is too large to be swallowed at a single gulp, and then some astonishing tactics are introduced. As described by Mr. Lankester, "A bite was taken from the body,

Fig. 578. — Pouch of frigate-bird.

being torn away by a wringing motion of the head which sent the carcass whirling, while the bird masticated the morsel into shape for swallowing.

Of course the fish began to obey the law of falling bodies. — and the bird, folding its wings tightly upon its body, dropped swiftly after it. The part bitten off being disposed of, another swoop downwards was made, the fish seized, and the upward swing repeated; and this process continued until the entire carcass was devoured. At the time of this visit these frigate-

Fig. 379. — White pelican (*Pelecanus onocrotalus*).

birds were oblivious of man's presence; and I was fortunate to secure this one by a well-directed shot. It measured eleven feet in alar dimensions, and weighed eight pounds."

In the frigate-birds, as our cut shows, there is a well-developed pouch beneath the throat; but this is far inferior to that of the next group to be mentioned, — the pelicans. These fish-eating birds are mostly found in the warm regions, and are too well known to need description. We have two species, a white and a brown, the former of which can scarcely be distinguished from the European form figured. Our white pelican breeds in the north, always selecting some remote and inaccessible spot for its nest. In the winter it goes south and stays along the shore of the Gulf states. In fishing, it swims along, striking the surface of the water with its wings, and scooping the fish into that admirable fish-net, the pouch under the lower jaw. At times this is so filled that it hangs nearly to the ground; and then the fish retires to the shore to eat them at his leisure.

"It is a pleasant sight," says Mr. Gosse, "to see a flock of pelicans fishing. A dozen or more are flying on heavy, flagging wing over the sea, the long neck doubled on the back, so that the beak seems to protrude from the breast. Suddenly a little ruffling of the water arrests their attention; and, with wings half closed, down each plunges with a resounding plash, and in an instant emerges with a fish. The beak is held aloft, a snap or two is made, the huge pouch is seen for a moment or two distended, then collapses as before; and heavily the bird rises to wing, and again beats over the surface with its fellows. It is worthy of observation that the pelican invariably performs a somersault under the surface; for, descending, as he always does, diagonally, not perpendicularly, the head emerges looking in the opposite direction to that in which it was looking before."

The common gannet is a northern species, which in winter descends on the more southern coasts. It is a beautiful white bird, with a long, gracefully curved neck, long, strong wings tipped with black. Its principal breeding places on our coasts are the islands in the Gulf of St. Lawrence. On Gannet Rock it is estimated that fifty thousand pairs of females come every spring to lay their eggs. They feed upon fishes, which they catch by plunging from on high, and it is said that each fish caught is swallowed at once, and those necessary to feed the young are disgorged after the return to the shore. The booby, the first-cousin to the gannet, is a darker bird, with brown predominating on the upper surface. It lives in the countries around the Gulf of Mexico, and usually ranges but a short distance north.

The bill of the cormorants, or shags, is much like that of the frigate-bird already figured, and they have also a pouch beneath the neck, but in several structural characters they differ considerably. Of the nearly forty species known but few need mention, but there is one which must not be neglected.

When Steller, a Russian explorer, was wrecked on Bering Island in 1741, he found the bird now known as Pallas's cormorant very abundant there. Indeed, it lived upon that island nearly a hundred years later. To-day the chances are that there is not a single individual alive. It has doubtless been exterminated at the hands of man. When speaking of the great auk, reference was made to its rarity in collections; but the present form is much more rare, for in all the museums of Europe there are but three or four specimens, and not a single one in the United States. It was a stupid bird, and furnished the natives with a large part of their meat during the long winter months.

On our eastern coast the common cormorant descends from the north to the middle states in winter; but in the summer it retires again to Labrador and the adjacent rocky islands, where on almost inaccessible cliffs it builds its nest and rears its young. A cormorant is proverbial for its voracity; it would seem sometimes as if it never could fill itself. Indeed, the word cormorant has become a synonym for rapacity and gluttony.

The story has often been told, denied, and reaffirmed, that in China cormorants are taught to fish for their masters. Now there remains not a doubt upon the subject; for it has been fully authenticated, though not with all the details which frequently surround it. They are trained in much the same way that one breaks in a retriever. At first they have a string tied to one leg, and are allowed to pick up fish thrown in, and are then drawn to the shore by means of the string, a peculiar whistle-call being made. Thus after a time they learn to come whenever the whistle is sounded. They are always rewarded by a portion of fish when they return to the shore. When fully trained, the fisherman takes them with him and, arrived at the spot, he points to the one which he wishes to dive first. Those who have witnessed the scene describe it as highly entertaining. The chosen bird dives into the water, while the row sitting on the edge of the boat preserve the utmost gravity. Soon the diver reappears, perchance with a fish in its beak, and brings it to the boat. If he should be unsuccessful, he dives again and again, and then if he catches no fish, he is in disgrace and is made to sit all alone. Now all the other birds are in a perfect flutter of excitement, just as though each were teasing to be allowed to show his skill. The fisherman with his long bamboo designates another bird, who immediately dives, while all the rest resume their customary gravity. To prevent the birds swallowing the fish they catch, a ring is placed around the neck. After the fishing is over for the day, each bird is rewarded with a quantity of food.

One more story, rather more apochryphal, must be told in connection with the cormorants. In times long gone by, the cormorant was a dealer in wool, and had for partners a brier and a bat. They sent their cargo by sea, and it was lost in the waves, which caused the firm to become bankrupt, — a disaster which affected the three partners in different ways, as may be noticed to the present time. The bat keeps secluded until night, so that he may avoid the creditors; the brier tries to make good the loss by pulling wool from every passing sheep: while the cormorant is continually diving into the sea in hopes that he may find the lost cargo. It scarcely needs to be said that this story belongs to the department of unnatural history.

Fig. 380.— Nest of the cormorant (*Phalacocorax bicristatus*).

In the snake-birds, or darters, the neck reaches its extreme development in length, and, as one watches a tropical pool inhabited by these birds, the appropriateness of the first-mentioned name is very apparent. While fishing, almost the entire body is submerged, while the long neck, small head, and long beak, as they rise from the water, have every appearance of a large snake. There are four species of snake-birds, — one from America, which occurs from Florida south to Brazil; one in Asia; one in Africa; and the fourth in Australia. All are fishers, and their abilities as divers make it difficult to shoot them. They build their bulky nests of sticks, moss, roots, and leaves, in trees, and lay three or four light blue eggs, coated with a chalky substance.

Unlike most of their relatives, the darters shun the sea-coast, but live in the dense swamps of warm countries. They are very timid, and are easily alarmed. When frightened they drop noiselessly into the water, and swim away to a considerable distance beneath the surface.

Last of the swimming birds comes the series typified by our common ducks, geese, and swans, the general appearance of which is familiar to all, and needs no description. One is never in a doubt concerning them. There is, however, one feature in their structure which needs mention. As is well known, most of these forms obtain a large proportion of their food from the ooze and mud at the bottom of the ponds and streams, and to aid them in this, most have a peculiar structure of the bill, which enables it to act as a sieve. On the inside of the mouth are a number of plate-like teeth, which at the same time serve to hold the ooze, and also to strain out the water when the head is raised. The whole operation can easily be seen when one of these birds is feeding. The geese, however, feed to a considerable extent on grass, and in these forms these plates are modified for nipping off the grass. These teeth are shown in our figure of the shoveller-duck, with which we open our account of the different species. It is, so far as color goes, a beautiful species, and is common to Europe, America, Asia, and Australia, but is less abundant with us than it is in the Old World. It is well named; for its bill, broadened at the end, is a true shovel, and is used as such in stirring up the mud at the bottom of the water, in the search for food. The sifting lamellæ, too, are developed to a greater extent than in most forms.

The mallard, distributed over nearly the whole world, needs no description, as it is so well known to all, either in its wild state or in its domesticated condition. The black duck of the hunters, the dusky duck of books,

Fig. 381.— Mallard ducks (*Anas boschas*).

is closely related, and in the eastern states, the most abundant of our ducks. It has a fine flesh, and is possibly the best of all the ducks for food. In its migrations — for it is to a large extent a migratory duck — a peculiarity is noticeable in the short distances which they travel; thus those which spend the winter in New Jersey go to Massachusetts for the summer, while those of the latter state migrate to Maine at the same time. It is a curious parallel to regular summer migrations of the human species. September and October are the great times for duck-hunting, and then the meadows and marshes resound with the bang of the sportsman's gun. To

FIG. 382. — Shoveller-duck (*Spatula clypeata*).

bring the ducks within range, decoys are used. Best for this purpose are three or four tame ducks, 'anchored,' one might say, in some small pond or stream; but in default wooden decoys, painted black and white, are used. The wild ducks see these, and, impelled by a feeling of sociability, soon settle down upon the water near them. This is the sportsman's opportunity. Concealed in shelter of boughs or grass, he opens fire with his double-barreled gun, and frequently has an opportunity to get in four shots before the alarmed flock are out of range. Many consider this 'sport,' but to others it seems but a refined cruelty.

The pintail duck is found in the northern portions of both continents, and derives its name from the long and pointed tail-feathers. It is far more abundant on the rivers of the interior than along the coast, and it goes to the far north, in the summer, to breed. Among the other river-ducks are the widgeon, the various species of teal, and the most beautiful of all the ducks,

FIG. 383.— Wood-duck (*Aix sponsa*).

— the wood, or summer, duck. No description, no woodcut, will do justice to its singular colors; white, metallic green, purple, chestnut, are all displayed upon its feathers, and in such a manner that Linnæus was fully justified in giving it the name he did; for *sponsa* means bridal, and certainly no bride was ever bedecked in a more beautiful or a better-fitting suit than is this duck. It is comparatively common throughout the United States, and is remarkable in its nesting habits. While most of the ducks seek the sedgy margins of ponds and streams, to build their nests, this species betakes itself to some hollow tree to lay its eggs and rear its young. It deems itself fortunate if it can find some deserted hole of the larger woodpeckers.

FIG. 384.— Canvas-back (*Fuligula vallisneria*).

The sea-ducks are separated by many characters, internal and external, from the river-ducks already mentioned. Possibly the most celebrated of all their numbers is the canvas-back, that delight of epicures. Like many of the other ducks it breeds in the north, and in October it returns to the

United States, and is especially abundant about the Chesapeake. Here, where the Vallisneria grows, they congregate in large numbers, and on the roots of this plant they feed. This plant is said to give the flesh its delicious flavor. Every possible method is used in hunting them; decoys of all sorts are employed; the gunning punts are covered so that the birds will not suspect their nature. The sport is said to be very exciting.

A peculiar interest surrounds the Labrador duck, an interest of the same nature as that surrounding the great auk, Pallas's cormorant, the dodo, and several other forms. In early years it was comparatively common on our eastern coast from Labrador south to Virginia. In the first half of the century it frequently appeared in Boston market. To-day it is nearly or quite extinct, the last known specimens having been taken near Elmira, N.Y., December 12, 1878. It is an extremely rare bird in collections, only thirty-four specimens being known (half as many as of the great auk), and of these about twenty are in American museums.

Fig. 385. — Labrador duck (*Camptolaimus labradoricus*).

What can have caused the extinction of this species is a question. In the case of the great auk, the dodo, and the flightless birds of New Zealand it is easy to explain their disappearance. These forms could not fly, and hence the persecutions of man rapidly exterminated them. With the Labrador duck the case was different. This bird was a strong flier, and made its annual migrations like most of its relatives. Various explanations have been suggested: some think that some epidemic disease may have been the cause, but in the opinion of the writer the explanation of Dr. Stejneger is the most satisfactory as far as it goes. What caused the first diminution in numbers it is impossible to say, but when once they became reduced beyond a certain number, their extinction is rapid. In their migrations there will be fewer, and with this the chance of losing the way will greatly increase, while the breeding places will become more disconnected, thus interfering with the formation of flocks, — an important element in the life of any migratory species. It may be that a few of these ducks are still in existence, but a few years more will certainly see the last of them.

The eider-ducks do not seem to be in any danger of immediate extermination; for in those northern countries where they are most abundant

Fig. 395. — Eider-duck (*Somateria mollissima*).

they are carefully protected. In the winter it comes as far south as New England, but its proper home is in Newfoundland, Labrador, Greenland, Iceland, and Norway. Here it builds its nest on the ground, constructing it of moss, grass, and sea-weed, and then lining it with the soft down which the mother plucks from her own breast. This down is highly prized, and in the north gives this bird a considerable commercial importance. At the proper time the down is collected and packed for export to Europe and America. It takes twelve nests to furnish a pound of down, and yet Greenland and Iceland each year export about three tons, the product of over seventy thousand nests.

A strange duck is the loggerhead, racehorse, or sidewheel-duck of southern South America. It is a strong and swift swimmer, using its webbed feet and wings in a peculiar manner, and from this fact derives its common name. It is also a splendid diver, but, strange to say, when it

Fig. 387. — Merganser, goosander (*Mergus merganser*).

becomes adult, it no longer flies, but depends upon its other powers to escape from danger.

Passing by the coots, scoters, and surf-ducks, we can only mention one more group of ducks, the mergansers, or sheldrakes, all of which have a more or less conspicuous crest on the crown of the head, while the lamellæ of the bill, instead of being used as sifting or straining organs, are converted into strong teeth well adapted to hold the fishes upon which these forms feed. Some make their nests by gathering leaves, straw, sedges, and the like into a large and high pile, while others prefer a hollow stump or the branches of some tall tree.

In the geese the sifting-teeth are modified for grazing; and the animals are far less clumsy while wandering about on the ground. Only a few of the many forms can be mentioned.

First comes the gray lag-goose of northern Europe. which is important as furnishing all our domesticated geese. Closely related to it is the American snow-goose which is dispersed all over our territory. though rare in the eastern states. Like its relatives of the poultry-yards it lives upon grass. at least when in the United States. It wanders through the fields. turning its head to the one side or the other. grasping a mouthful and tearing it off with a quick jerk of the neck. Its breeding-place is far to the north in the British possessions. It leaves for the north early in the spring. and returns late in the fall.

Better known. or at least more familiar. is the common wild goose. — the Canada goose of the books. In the early spring. or late in the fall, their familiar 'honk' is heard ; and on looking up one sees the flock passing north or south. It flies in a V-shaped body. an old gander at the apex. while on either limb are a varying number of birds. all winging their way in a heavy. laborious manner. Now one will cross from one side to the other of the flock. but why no one knows. Again the leader sounds his note. and a responsive series of 'honks' comes from the rest. When the weather is fair. their course is straight. and their elevation is not great : but

Fig. 388. — Common wild goose (*Bronta canadensis*).

they continue for long distances. alighting occasionally to rest on the waters of some secluded pond. When the weather is foggy. they are all mixed up. The oft-extolled 'instinct' fails them : and they know not which way to turn.

The barnacle-goose of Europe has been found but two or three times in the United States. It is particularly interesting from the fact that it is the species which was said to breed from barnacles, allusion to which was made on a preceding page. Its name in the older spelling was bernacle (a derivative from Hibernia, Ireland); and so similar was this to barnacle that the myth arose in all probability from a confusion of names.

The swans, with their long and gracefully curved necks, are in many respects intermediate between the ducks and the geese. Many of them are semi-domesticated, and are kept on the lakes and ponds of parks, where their beautiful shapes and white wings make them attractive objects. Best known of all are the mute swans of Europe, with dazzling white plumage and scarlet bills. As they glide smoothly along, their necks curved like the letter S, and their wings half expanded, they are worthy of all the words used by poets.

Our two swans — the whistling and the trumpeter swans — are much alike, each having a peculiar arrangement of the wind-pipe, something like that of the cranes, to be mentioned farther on, which gives the voice the peculiar character implied by the common names. These are migratory birds, breeding mostly in the far north, and coming into the United States in

FIG. 389. — Whistling-swan (*Cygnus columbianus*).

winter. The trumpeter is much the larger of the two, and sometimes weighs nearly forty pounds.

All swans, however, are not white. In South America occurs a species with a black or dark brown head and neck; while Australia, that land of wonders, furnishes swans entirely black, except the white on the wings and the red and white on the bill. These black swans have been introduced into parks along with the common European species.

Among the more distant relatives of the ducks and geese is the curious horned screamer, or unicorn bird, of South America. In their general appearance they are more like the wading birds. Their legs are long, their feet not webbed; but their internal anatomy settles the question. The best-known species looks something like a rail; but its wings are furnished with strong spurs, like those of the jaçana, to be mentioned shortly, while from the top of the head arises a long and slender horn which curves gracefully forward, giving the bird the most unique appearance. Its voice is described as a harsh scream like the bray of a jackass. It is, however, excelled in vocal powers by a related species from Buenos Ayres. Of it Mr. Gibson says, "It is much addicted to soaring, and scores may be seen at a time, rising in great spiral circles till they become mere specks in the sky, and actually disappear at last. Even at this elevation the cry is distinctly audible, and has often drawn my attention to the bird as having really vanished into the blue ether. The cry, which may be often heard at night, is frequently indulged in, and consists of the syllables *cha-ha*, uttered by the male, while the female invariably responds to it, or rather follows it up with *cha-ha-li*, placing the accent on the last syllable. Preparatory to producing it, if on the ground, the bird draws back its head and neck slightly; and at that moment, if one is sufficiently near, the inhalation of air into the chest may be faintly heard. The note is of great strength and volume, and is still distinguishable a couple of miles away, if the day should be calm."

WADING BIRDS.

The flamingoes are strange, but in another way. They look like swans on stilts, all but the beak, and that is most peculiar. At the middle it is bent abruptly downwards, the upper half shutting into the lower. The legs are long, and terminate in webbed feet; while the neck is extremely long, and the bill is provided with sifting-plates, like those of the ducks. As will be seen, these features (besides many others anatomical) make it a form intermediate between the swimming and the wading birds, some placing it in the one group, while others assign it a position in the other.

Many of the habits of the flamingoes are well known, but others were not settled until a very recent date. A group of them feeding is a beautiful sight, and it is no wonder that the inhabitants of Spanish America, noticing their lengthened lines, their subjection to a leader, and their red plumage, call them 'English soldiers.' Mr. Richard Hill described, many years ago, their habits in confinement. "I was struck," he says, "with their attitudes, with the excellent adaptation of their twofold character of waders and swimmers, to their habits, while standing and feeding in the sort of shoal which we made them in a large tub upon deck. We were here able to observe their natural gait and action. With a fine exactness, like a man treading a wine-press, they trod and stirred the mashed biscuits and junked fish with which we fed them; and plied their long, lithe necks, scooping with their heads reversed and bent inwardly towards their trampling feet. The bill being crooked and flattened for accommodation to this reversed mode of feeding, when the head is thrust down into the mud-shoals and the sand-drifts, the upper bill alone touches the ground."

FIG. 390. — Flamingo (*Phœnicopterus antiquorum*).

So much has long been known, but about their nesting habits there existed great uncertainty until the present decade. The nests — huge hillocks of clay — had long been known; and as some of these were very high, the old account of their nesting was all but universally accepted. According to this, the birds sat astride the nest, like a man on horseback, the long legs hanging down on either side. Time went on, and, so wary were the birds, no one had a chance until 1881 to settle the question whether the birds did really sit in this unbird-like posture. In that year (and the observation has since been confirmed) it was definitely ascertained that the old idea was wrong. One account says: "We approached within some seventy yards before their sentries showed signs of alarm, and at that distance with the glass observed the setting birds as distinctly as one need wish. Their long red legs doubled under their bodies, the knees projecting as far as or beyond the tail, and their graceful necks neatly curled away among their back-feathers, like a sitting swan, with their heads resting on their breasts,—all these points were unmistakable. Indeed, it is hardly necessary to point out that in the great majority of cases (the nests being hardly raised above the level of the hard mud), no other position was possible." One correction needs to be made in the foregoing quotation; the "knees" must have been the heels.

The rails and their allies are familiar to all. They are long-legged birds which frequent marshy and swampy places. During the day they largely keep themselves concealed, and it is as evening comes on that their loud, harsh voice is most heard. They wend their way between the rushes with the utmost ease, and prefer their legs to their wings to escape from the hunter and his dog. Dr. Coues thus describes the Virginia rail as he saw, or rather heard it, at the Sink of the Mojave: "At nightfall some mallard and teal settled into the

Fig. 391.—Carolina rail (*Porzana carolina*).

rushes, gabbling curious vespers as they went to rest. A few marsh-wrens had appeared on the edge of the reeds, queerly balancing themselves on the thread-like leaves, seesawing to their own quaint music. Then they were hushed, and as darkness settled down, the dull, heavy croaking of the frogs played bass to the shrill falsetto of the insects. Suddenly they, too, were hushed in turn, frightened, it may be, into silence; and from the heart of the bulrushes, '*crik-crik-rik-k-k*' lustily shouted some wide-awake rail, to be answered by another and another, until the reeds resounded. Then all was silent again till the most courageous frog renewed his pipes. The rails are, partially at least, nocturnal. During such moonlight nights as this they are on the alert, patrolling the marshes through the countless covered ways among the reeds, stopping to cry 'all's well' as they pass on, or to answer the challenge of a distant watchman."

The gallinules and purple gallinules must also be mentioned among the rails: our common purple gallinule of the south being a beautiful bird, with bluish green and purple plumage, and very long toes; and the true gallinule of the United States, of a black or brown color, with a scarlet shield on the top of the bill. Here, too, must be mentioned a huge rail-like bird, the geant, which, two hundred years ago, lived on the island of Rodriguez, along with so many other now extinct birds. It was six feet in height, with white plumage.

The coots are the most thoroughly aquatic of any of the rail-like forms, inhabiting reedy marshes, swimming with ease in the reaches of water, and rarely taking to their wings to escape from danger. They are shy birds, retiring to the reeds when disturbed, and there they build their nests. Sometimes these structures rest on the soil, and sometimes they are rafts of reed-stems, moored among the rushes, much like the nests of the grebes.

The cranes are large birds, which each year, like the ducks and geese, travel to the north to breed, and then, in the fall, return to the south to spend the winter. They fly in the same V-shaped flocks as do the wild geese, and the loud trumpeting of some of the species

Fig. 302. — Geant (*Leguatia gigantea*), as restored by Schlegel.

attracts the attention as the flocks pass overhead. This note is loudest in the whooping-crane, and the apparatus for producing it is interesting. In most birds the windpipe has a straight course from the throat to the lungs: but in the cranes and some others it is variously coiled, the extreme being found in the whooping-crane. Here the windpipe enters the breast-bone, and coils round and round, the whole, both in appearance and effect, resembling the convolutions of the horns used in our brass bands.

Seen in flight, the cranes seem heavy; their wings move slowly and laboriously, and one watching them thinks that their volant powers are about exhausted; and yet they go on, a regular game of 'follow the leader,' turning here and there, with the course of the stream they are following. One watches them as they go, thinking every minute that they must light; but now the individual birds can no longer be distinguished and shortly the whole flock is out of sight.

Judge J. D. Caton kept three sandhill-cranes in his grounds for several years, and interesting pets they made. They could fly well, but they never made any attempts to escape, even at the migratory season. They "could often be induced to dance and play with me in their awkward, but very amusing, way. They are inclined to be imitative. Forty years ago, when they were very abundant in this country, a farmer whom I

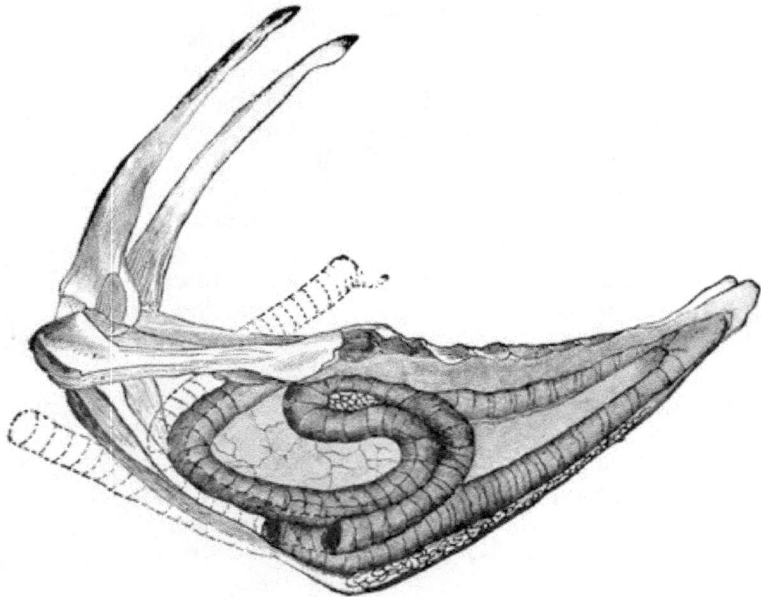

Fig. 393. — Convolutions of the windpipe in the breast-bone of the whooping-crane.

well know, assured me that he had one in domestication which, when a year old, would fly on to the haystack and tramp around in imitation of the boy, and would also take the lines in its beak and follow the horses, breaking prairie, with a stately strut that was very amusing." One of the birds left its fellow and took up with the pigs for companions, following them with the most constant devotion. One of the females laid four eggs, and the male "spent most of his time pretty near the nest, and guarded it with great fidelity, and defended it with courage. If a cow or deer came near it, he flew at it in a rage, and a few thrusts with his sharp beak sent it away in a hurry; and if he saw a buggy coming in that direction, he raised his coarse, harsh voice in so threatening a way as not to be mistaken; and if it came too near, he flew at it, attacking either the buggy or the horse, whichever happened to be nearest; and if it went within, say fifteen or twenty feet of the nest, the female would leave the eggs and join in the attack, and the premises were soon cleared. . . . In fact, he was almost as constant in his watchfulness, and as pugnacious in his conduct, as a wild (Canada) gander, whose goose was sitting across the ravine. It was the habit of this cock, whenever the hen left the nest to seek for food, to take her place; but he cut an awkward figure, sitting on the nest, for his long legs seemed to be much in the way, while the female had managed to assume a rather graceful position while performing that maternal duty."

The bustards of the Old World — some smaller than our domestic fowl, others as large as a turkey — are hardly 'waders' in the literal (or litoral) sense of the word; for they are birds of the plains. They are strong runners, and some prefer their legs to their wings to escape from danger. This fact formerly led to their union with the ostriches in a group of runners; but, still, in every point of their anatomy they are wading birds. A singular feature in their structure is the pouch — enormous in some species, absent in others — below the neck, which during the mating season is inflated with air from the throat, as a part of the 'showing off.' One species, the houbara of India, is regarded as a game-bird, and is hunted with camels, the hunter circling round and round in a spiral, each turn bringing him nearer the game, until at last it is within gunshot. But as the spiral grows smaller, the birds suddenly disappear. They have squatted flat upon the sand, and so perfectly do their colors harmonize with the surroundings that they are absolutely invisible. "You will have a bird rise suddenly, apparently out of the earth, within five yards of you, from a spot where there is not a blade of cover, and on which your eyes have perhaps been fixed for five seconds." The vertical sun, casting no shadow, aids in the concealment.

The herons, egrets, bitterns, and boat-bills form a group of birds, some members of which are found in every country. Most voracious birds are they. Other animals after gorging themselves require a period of rest; not so the herons. They suffer with no dyspepsia, but seem to digest their food almost as soon as it is swallowed, and are then as hungry as before. Mr. Endicott's experience with a tame specimen of the night-heron may be cited as illustrative. The heron had been trying to catch the chickens, and so Mr. Endicott "took to fishing for him, and then to my sorrow, I found out what a heron's appetite is; and thought, with pity, of the poor parent birds in the swamp with six or eight such maws to fill. Five bream, as large as my hand, were not too much of a meal for him. He would catch them, all alive, out of the tub of water, by the middle of the back, toss them up until he got them in the right position, head-first down his throat; then he would swallow them by dint of great exertion, his neck presenting a curious appearance, as the fish, four inches broad, passed slowly down, making occasional convulsive attempts to struggle; a proceeding which seemed to enhance the pleasure of the bird."

We pass our herons and egrets by with mere mention, — the great blue heron widely distributed over the United States, the great and the little white egrets, and their closely similar European counterparts, the green heron, and the night-heron already mentioned; for we wish room to make a quotation from the pages of Mr. W. H. Hudson regarding one of the smallest of the South American bitterns. "One day in November, 1870, when out shooting, I noticed a little heron stealing off quickly through a bed of rushes, thirty or forty yards from me; he was a foot or so above the ground, and went so rapidly that he appeared to glide through the rushes without touching them. I fired, but afterwards ascertained that in my hurry I missed my aim. The bird, however, disappeared at the report; and thinking I had killed

FIG. 394. — Great blue heron (*Ardea herodias*).

him, I went to the spot. It was a small, isolated bed of rushes I had seen him in; the mud below and for some distance round was quite bare and hard, so that it would have been impossible for the bird to escape without being perceived; and yet, dead or alive, he was not to be found. After vainly searching and re-searching through the rushes for a quarter of an hour, I gave over the quest in great disgust and bewilderment, and, after reloading, was just turning to go, when, behold! there stood my heron as a reed, not more than eight inches from, and on a level with, my knees. He was perched, the body erect, and the point of the tail touching the reed grasped by the feet; the long, slender, tapering neck was held stiff, straight, and vertically; and the head and beak, instead of

FIG. 385. — Little white egret of Europe (*Garzetta nivea*).

being carried obliquely, were also pointing up. There was not, from the feet to the tip of the beak, a perceptible curve or inequality, but the whole was the figure (the exact counterpart) of a straight, tapering rush; the loose plumage arranged to fill inequalities, the wings pressed into the hollow sides, made it impossible to see where the body ended and the neck began, or to distinguish head from neck, or beak from head. This was, of course, a front view; and the entire under surface of the bird was thus displayed, all of a uniform dull yellow like that of a faded rush. I regarded the bird wonderingly for some time; but not the least motion did it make. I thought it was wounded or paralyzed with fear, and, placing my hand on the point of its beak, forced the head down till it touched the back; when I withdrew my hand, up flew the head, like a steel spring, to its first position. I repeated the experiment many times with the same result, the very eyes of the bird appearing all the time rigid and unwinking like those of a creature in a fit. What wonder that it is so difficult — almost impossible — to discover the bird in such an attitude! But how happened it that while repeatedly walking round the bird through the rushes I had not caught sight of the striped back and the broad, dark-colored sides? I asked myself this question, and stepped round to get a side view, when, *mirabile dictu*, I could still see nothing but the rush-like front of the bird! His motions on the perch, as he turned slowly or quickly round, still keeping the edge of the blade-like body before me, corresponded so exactly with my own that I almost doubted that I had moved at all. No sooner had I seen the finishing part of this marvellous instinct of self-preservation (this last act making the whole entire) than such a degree of delight and admiration possessed me as I have never before experienced during my researches, much as I have conversed with wild animals in the wilderness, and many and perfect as are the instances of adaptation I have witnessed."

The bitterns are celebrated for their vocal performances. The Latin name *Botaurus* means bull-voiced, and all the other names — stake-driver, mire-drum, bull-of-the-bog, butter-bump — have reference to the strange 'booming' note of our or the European species. In the Old World many a superstition surrounds the bittern; but with us no such old wives' tales have obtained currency. Then, again, many a strange story was told of the way in which the bittern makes its note. Some thought the bird put his bill in a hollow reed,

Fig. 36. — Bittern, or stake-driver (*Botaurus lentiginosus*).

while others claimed he stuck it in the mire. Chaucer says: —

"And as a bitore bumbleth in the mire
She laid hire mouth into the water doun."

The European species makes a roar which may be sometimes heard for three or four miles; and in making it the bird " lifted the head, threw it backwards, put it again rapidly into the water, producing a roar that startled me." The American bittern is not so loud-voiced as his European relative; and the character of his note is well described by his common name stake-driver. "I have often," says Mr. Samuels, "when in the forests of northern Maine, been deceived by this note into believing that some woodman or settler was in my neighborhood, and discovered my mistake only after toiling through swamp and morass for perhaps half a mile." The note, as the writer has often heard it, sounds almost exactly like the stroke of a mallet on a fence-post; yet, strange as it is, it is a love-song, and doubtless sounds far more sweet to Mrs. Bittern than would any lay of Herrick or Suckling.

The shoe-bills and boat-bills are aberrant relatives of the herons, but are but remotely allied to each other. In each the bill is very broad, and has a remote resemblance to the objects embraced in its name. The shoe-bills are African, while the boat-bills, one of which is represented in the plate of the anaconda, opposite page 374, are South American.

The group of storks are hardly represented in the forms of the United States; but in South America there are some, while in the Old World they play an important role. There dwell the curious adjutants, "whose thin long legs," as Mr. Forbes expresses it, "always suggested the idea that they had escaped from some taxidermist's hands when he had just got the length of running the wires up their shanks." These birds furnish the valuable marabou plumes, once so highly esteemed as ornaments. They are scavengers; and in some parts of the East Indies they are protected by law.

Then there is the familiar stork, which each winter goes to Africa, returning in the spring to make her nest on the roof of some German dwelling, and bringing all sorts of luck and good fortune with it. Year after year as the trees open their leaves, back come the same pair of storks to the roof-tree; and soon the callow brood is hatched. In their flight they closely resemble the flocks of wild geese in this country.

The wood-ibis is our only common member of the stork family in the United States; and this does not wander north of the southern states. There, however, it is common; and Audubon speaks of flocks of " several thousands," though now no such numbers are seen. They breed abundantly in Florida as well as in the southwestern territories. Dr. Coues, among others, has written an interesting account of their habits and their appearance in flight, from which we make some abstracts. The carriage is firm and sedate; each leg is slowly lifted and planted with deliberate precision. when the birds walk unsuspicious of danger. When taking wing, it springs powerfully from the ground, bending low to gather strength, and for a little distance flaps hurriedly with dangling legs, as if it was much exertion to lift so heavy a body. But fairly on wing, clear of all obstacles, the flight is firm, strong, and direct, performed

FIG. 397. — Adjutant, or marabou stork (*Leptoptilos crumenifer*).

with continuous, moderately rapid beats of the wing except when the birds are sailing in circles; when, with wide-spread, motionless pinions, they go round and round as if supported by magic. A score or more cross each other's paths in interminable spirals; their snowy bodies tipped at the wing-points with jetty black. clear cut against the sky; they become specks in the air, and finally pass from view.

This soaring of birds is an oft-discussed problem, and it may be as well to mention it here as anywhere else. Familiar as we are with the law of gravitation, it seems a strange thing that a bird, without a single stroke of the wing, after the first momentum is gained, should be able to go up, up, up, until clear out of sight. We would rather expect it to tumble to the earth as soon as the first impulse was lost. Many are the solutions that have been offered, but there is not one that is wholly satisfactory. The best advanced is briefly as follows: First, the bird, in mounting thus, with rigid pinions, always moves in a spiral; and second, the soaring never takes place unless there is at least a light breeze. The bird, either by flapping his wings, or start-

Fig. 398. — Wood-ibis (*Tantalus loculator*).

ing from some elevation, gets his initial momentum, and then the soaring begins. Round and round he goes, each complete turn bringing him higher, but at the same time he has drifted away with the wind. As he circles round, at one time his course is with the wind, at another it is directly in the face of it. When going with the wind he allows his body to fall, but holds his wings in such a position that the downward motion is converted into forward speed; then round the turn he goes and faces the wind, and then, like a kite, the breeze carries him up, and to a higher point than on the previous turn. Thus it goes on, turn after turn, there being a slight drifting to leeward, and a progress upwards, like the frog in the well, a gain of three feet and a loss of two, the result being in each cycle a positive gain; and at last the bird is out of sight.

FIG. 389. — European spoonbill (*Platalea leucorodia*).

Of the true ibises none has more celebrity than the sacred ibis of ancient Egypt. There has been considerable confusion as to just what species the sacred ibis was. To-day travelers going up the Nile are shown a large, buff-backed heron, and are told that that is the bird in question. Not so. The true ibis rarely appears north of Khartoom, and some think the old Egyptians imported the bird from the south. However this may be, there is no question as to the reverence shown for it. It was ubiquitous, and a regular nuisance, although it did some good as a scavenger. And when it died, its body was embalmed and placed in the tombs.

In the United States we have several species of ibises, but only one, the white ibis, is at all abundant, and this is confined to the southern states. Rarely the scarlet ibis of South America enters our limits, as also does the jabiru. The scarlet ibis is one of the most beautiful of birds, and is a frequent inhabitant of zoological gardens; but in confinement the feathers lose their beautiful scarlet hue and fade away to a rosy tint, scarcely less beautiful. The spoonbills are merely modified ibises, in which the cylindrical, regularly curved bill is broadened and flattened, so that it bears a distant

FIG. 400. — Woodcock (*Philohela minor*).

resemblance to a spoon. The European species figured is almost entirely pure white, but our American species has the wing-coverts a bright red.

The snipe family is a large one; it embraces many species of great interest to the sportsman, for here come the snipe and woodcock, the doebirds and curlews, and many others, some of which will be mentioned. In all the bill is delicate, and is richly supplied with nerves, and thus plays an important part in the search for food. In their hunting for means of subsistence these birds probe the moist earth with the bill, and then the sensitive surface comes into play; for it aids in recognizing the presence of the worms, etc., on which they feed. These 'borings' are shown in the adjacent cut of the woodcock; but they subserve a purpose not contemplated in the economy of these birds, for they are sure guides to the sportsman in his search after game.

The woodcock and the snipe may be regarded as the typical members of the group, and hence will be treated first. Neither need any introduction, as both are pretty well distributed over the United States. In the southern states the woodcock is resident the year round, but in the north, though a few spend the winter, it is mostly a summer resident. It is largely a nocturnal bird, and at nightfall it saunters out from the swamps and woods and begins its real life. It mates at dusk, and in the early evening the pairs may be seen in the air, going through the most eccentric motions, mounting up and then suddenly darting down to earth. At night it also searches for its food, and its peculiar note, chip-per, chip-per, chip, or some of its 'bleating' modifications, are welcome sounds to the sportsman. From the gamy point of view the woodcock acts exactly as it ought to act. When flushed by the dog it starts in the proper way, flies as it ought to fly, and tumbles, when shot, in the most gratifying manner. A change in habits would be of an advantage to the species, if not to the sportsman. If it would only refuse to fly, or to stand still, but take to the trees, or the dense underbrush, it would be much the better for it; but who can expect much of such a stupid-looking bird, with its eyes placed far back where no intelligent bird would have them? The snipe is an equal favorite with the sportsman, and its habits are much the same, the differences in flight being of such a character as to interest only the gunner. With all the excitement of the sportsman — the bagging of game only for the pleasure of bagging it; the wounding of birds, which escape to the swamps to die an agonizing death — the writer has not the slightest sympathy. It is nothing but cruelty, unmitigated cruelty. And yet men will go out, day after day, with gun and dog, and shoot birds so small that twenty of them would not make a decent meal, and call the occupation "sport."

The subject of our next cut is a strange-looking, snipe-like bird, which occasionally appears in the United States, but which is abundant in some parts of the Old World. The most remarkable feature is the ruff about the neck, from which the name of the bird is derived. This is only developed in the male bird at the breeding season, and seems to be of the nature of armor, and of great use in the combats in which these birds indulge in the breeding season. The ruff, unlike most of the snipe, is polygamous, and each spring the males fight for the possession of the females of the flock. Year after year they go to the same spot for these duels. Then the males fly at each other, striving to strike with the long pointed bills, and receiving the blows of the adversary on the collar of feathers, which is held erect. Soon one of the birds admits himself defeated, though usually he has suffered no injury. After the mating, the feathers of the ruff fall out.

Allied to the snipe are a long series of birds known under a host of names, — willets, sanderlings, sandpipers, knot, dotterel, and the like; birds with much the same habits, and which may be dismissed with this mere mention. Dr. Coues's account of one will answer for all.

"Fogs hang low and heavy over rock-girdled Labrador. Angry waves, palled with rage, exhaust themselves to encroach upon the stern shores,

FIG. 401. — Ruff (*Pavoncella pugnax*). One of their fights is represented in the background.

and baffled, slink back howling to the depths. Winds shriek as they course from crag to crag in mad career, till the humble mosses that clothe the rocks crouch lower still in fear. Overhead the sea-gulls scream as they winnow, and the murres, all silent, ply eager oars to escape the blast. What is there here to entice the steps of the delicate birds? Yet they have come, urged by resistless impulse, and have made a nest on the ground in some half-sheltered nook. The material was ready at hand in the mossy covering of the earth, and little care or thought was needed to fashion a little bunch into a little home. Four eggs are laid with the points together, so that they may take up less room, and be more warmly covered; there

is need of this, such large eggs belonging to so small a bird.   As we draw
near, the mother sees us, and nestles closer still over her treasures, quite
hiding them in the covering of her breast, and watches us with timid eyes,
all anxiety for the safety of what is dearer to her than her own life.   Her
mate stands motionless, but not unmoved, hard by, not venturing even to
chirp the note of sympathy and encouragement she loves to hear.   Alas!
hope fades and dies out, leaving only fear; there is no further concealment
— we are almost upon the nest; almost trodden upon she springs up with
a piteous cry, and flies a little distance, re-alighting, almost beside herself

FIG. 402. — Sanderling (*Calidris arenaria*).

with grief; for she knows too well what is to be feared at such a time.   If
there were hope for her that her nest were undiscovered, she might dis-
simulate, and try to entice us away by those touching deceits that maternal love inspires.   But we are actually bending over her treasures, and
deception would be in vain; her grief is too great to be witnessed unmoved,
still less portrayed; nor can we, deaf to her beseeching, change it into
despair.   We have seen and admired the house, — there is no excuse
for making it desolate; we have not so much as touched one of the
precious eggs, and will leave them to her renewed and patient care. . . .

" But except from ourselves, the birds have little to fear. Their enemies are few, they lead a merry, contented life, and it is no wonder they
increase and multiply till they become like armies as to numbers. Besides
being gregarious among themselves, they are sociable with other birds, and
there is hardly a gathering of waders of any sort anywhere that the peep
family is not represented in. Gadabouts, perhaps, they are, but no scandal-
mongers; ubiquitous, turning up everywhere when least expected, but
never looked ill upon; bustling little busybodies, but minding their own
business strictly. Besides environing a continent on three sides at least
(and perhaps on the Arctic shores as well), not a river, not a creek or
pond, the banks of which are not populated at one season or another; the
track of their tiny feet imprinted on the sand of the sea-shore and the soil
of the inland water shows where they have gone. Their numbers swell in
no small degree the great tide of birds that ceaselessly ebbs and flows once
a year in the direction of the polar star; they taken away, a feature of
the land would be lost. Altogether they became imposing, though singly
insignificant. If we do not know just what part is given out to them in
the great play of nature, at least, we may be assured that they have a
part that is faithfully and well performed."

In the curlews the long bill is bent downwards, in the stilts it is
straight, and in the avocets it is curved upwards. The avocets, farther,
have a hind toe which is lacking in the stilts. They are more aquatic
than some of the forms mentioned above. The American avocet is especially abundant in the Mississippi basin in summer, while the black-necked
stilt is more common along the southern shores. In habits these two are
much alike, but the stilt has been greatly maligned in the matter of its
slender legs, and it has even " been asserted that its leg-bones are as limber
as a leathern thong, and that they can be bent up without being broken."
This perfectly absurd statement doubtless had its origin in the exaggeration of some other statement as to its tottering attitude when the bird
first alights.

The plovers are well-known birds of migratory habits, breeding in the
far north, and stopping in our northern states only during their spring
and autumn migrations. One of the strangest members of the group is
the crooked-bill plover of New Zealand. In this the bill is bent to the
right in a curious manner. The first specimen was taken in 1833, and
then for years no other specimen was known, and, as a natural result, it
was regarded as a monstrosity. In 1869, however, another was found,
and since that time it has been definitely ascertained that this is the
normal condition of affairs, and that the crook of the bill is of use to the
bird. It seeks its food of insects and shells underneath stones, and here is

where it comes into play. With it it can probe beneath a stone with far greater ease than if the bill were straight.

Another of the plovers is the celebrated leech-eater, or trochilos of the Nile, which was and is still said to act as attendant to the crocodile, freeing it from parasites of all sorts. This story appears in the writers of antiquity, but with some variations. The bird was said to live at peace with the rep-

Fig. 403. — European avocet (*Recurvirostra avocetta*).

tile, and to go inside his mouth and pick off the leeches and pull the fragments of food from between his teeth. Next intervened a period of scepticism and doubt; but now in its essentials the old story has been confirmed: the bird picks the parasites from the body, and even ventures to snatch the morsels from between the teeth. It may be that some future student will again bring forward the story that the crocodile opens his mouth and invites the bird to enter the cavity, there to play the part of a living toothpick.

The turnstones and oyster-catchers are also aberrant plovers. The oyster-catcher is one of the wariest of the shore-birds, and one that never ventures far from the salt water. It has a strong, compressed bill, very useful in prying open the shells of the smaller mussels, and in cracking the bodies of the fiddler-crabs. The turnstone has somewhat similar habits, and as its name implies, wanders about the shore, turning over the stones in its search for food. The species figured has a nearly cosmopolitan distribution, being found on the sea-shores of nearly all countries.

Fig. 404. — Turnstone (*Arenaria interpres*).

The jaçanas, of which there are four species, are tropical birds which at first sight seem to have but little relationship to the plovers, but would ordinarily be regarded as rails. They have the same long legs and toes, and the general facies of a rail; but anatomy, which decides such questions, clearly indicates their affinities to be with the plovers. Our cut of a South American species will show the general appearance of the American forms, one of which just enters the United States in Texas. The long toes, terminating in the extra long and sharp claws, make it easy for them to run about on the leaves of water-plants, while the strong spurs on the wings as well as the claws are put to use in their combats. The species figured is a beautiful bird. the dark parts of the plumage being a rich, deep. purplish black. the lighter portions a copper-green.

The last of the wading birds to be mentioned is a curious form from the Antarctic Ocean. They have a sheath on the top of the bill from which they derive their book name of sheath-bills. The whalers call them

Fig. 405. — Jaçana (*Jacana spinosa*).

'white paddy' and 'sore-eyed pigeon.' the latter name having reference to the pale pink eyelids. The most studied form is the one which lives on Kerguelen Island. and the various expeditions which went to that desolate land some years ago to observe the transit of Venus. have told us about all we know of it. They are fearless birds. and exhibit a very curious disposition, and in many ways they remind one of the pigeons and grouse. with which. indeed. they were formerly grouped. Dr. J. H. Kidder has furnished some notes on their habits, from which we make a few abstracts; but in such a way that no quotation marks can be used. Specimens were captured by hand, all that was necessary being to stand perfectly still until they came within reach, and the first one caught served as a lure for others. They fought among themselves, using the bill, but not the spurs upon the wings. During the night they clucked and pecked at the woodwork, so that one would think that a flock of chickens were about. There were different opinions as to their edibility. The German observers thought them the best birds on the island; the whalers said they would " do very well when very short of fresh meat"; while the American party did not experiment on them at all, the flesh being very dark, and apparently very tough.

HOATZIN.

The South American hoatzin, or, as it is often called, the cigana, or gypsy, is a strange bird. It is dark brown, varied with reddish, and is about the size of a pheasant; indeed, until recently it was grouped with those forms. But a study of its anatomy reveals the fact that it is *sui generis*. No other bird has such a skeleton, none such an arrangement of muscles. Of its development nothing is known beyond the fact that the young birds have a couple of claws upon the wings, but these are not of such a character as to warrant the extravagant accounts which have been given of them. They are scarcely more prominent than the spur of the jaçana, though they are placed on two two-jointed digits. They are said to be used as aids in locomotion in clambering about in the trees, and are lost in later life.

It is a thoroughly arboreal species, living in small flocks of twenty or thirty individuals in the alluvial regions of northern South America. It never appears on the ground, and it is also said not to frequent high trees, but rather the lower bushes. It is polygamous, as are our domestic fowl, and its voice is a hard, grating hiss. It is nowhere domesticated; indeed, it would not be a desirable form; for its food, chiefly the leaves of one of the Arums, give the flesh a most disagreeable taste. Says Mr. Bates, " The flesh has an unpleasant odor of musk combined with wet hides — a smell called by the Brazilians catinga; it is therefore uneatable."

SCRATCHING BIRDS.

The common name, scratching birds, and its Latin equivalent, Rasores, have been applied to the group which is typified by our common barnyard fowl, the members of which obtain their food by scratching the earth.

This habit in itself is of but slight importance, but it is associated with so many structural features as to render the division a perfectly valid and natural one. It embraces between three and four hundred species.

First comes the group of Old-World quails of which the common migratory European species is best known. It is a small bird, only twice as large as the figure. It travels in large flocks when on its migrations, and as is the case with many other birds, its journeys are made at night. Sev-

Fig. 406. — European quail (*Coturnix dactylisonans*).

eral times the attempt has been made to introduce this valuable game-bird into America, and many hundreds have been turned loose, but the result has not been very successful.

The true partridges are also European and Asiatic, and differ considerably from the forms which we call by the same name. Some of these forms are very abundant, and furnish fine sport for the hunter; but an enumeration of them in the space which we can afford would prove but little more than a catalogue of names. One or two, however, deserve mention. Among these are the red-legged partridges, celebrated not so much for their gamy qualities or their value as food as for their extreme pugnacity. In olden times they were kept much as game-cocks are kept to-day, and their battles then excited as much interest among the higher classes as a cocking-main does among the so-called sporting fraternity of the present time.

Far better as game-birds are the thirty odd species of francolins, whose centre is the region around the Mediterranean. They are all noisy birds, with a sharp, quick note, which frequently informs the hunter of their whereabouts. The magnificent snow-pheasants of the high mountains of Central Asia should also be mentioned. As their name implies, they make their home near the snow-line, and they are so wary that it is only with difficulty that the hunter can get a shot at them.

Turning now to the American continent, we find a large number of forms; possibly not so great as those of the Old World, but certainly of far more interest to us. Between the partridges of the eastern and the western continents there are many differences, and the ornithologist who is shown a specimen, even if it be a form he has never seen before, can at once tell approximately whence it came. These distinctions, however, need not trouble us, for systematic details are something beyond our scope.

First of the American partridges come a few (about twenty) forms from Central and South America, of which the chivelua may serve as the type. It is thus described by Mr. Gaumer : —

"The chivelua is rare in the settled portions of Yucatan, but is common in the forests and beyond civilization. It is never found in fields nor in open lands, and does not live in flocks like the quails, but is generally found in pairs; and more rarely two or three pairs are found together. It lives on the ground, where it spends most of its time scratching about old logs and trees in search of insects, worms, and seeds, which form the principal part of its food.

"The numerous dusting-holes made by this bird are always sure indications of its presence; but owing to its habit of remaining immovable until all danger is past, it is very difficult to see it. In addition to this, the colors of the feathers are exactly similar to the lights and shades of dead leaves upon the ground. This bird is very tame, and in fact has no fear of man, and only flies when hard-pressed; and even then it never flies high nor far away. I have often stood for some minutes when among the dusting-holes of these birds, looking about the ground for them, and only after a long time and close watching I have been able to spy them out. In this way I have taken them many times in an insect net which had only a common walking-stick for a handle. It makes a beautiful pet and

becomes very tame, but does not live long in confinement. At nightfall
the chivelua sings his sweet evening song, beginning with a low whistle
which is three times repeated, each time with greater force; then follow
the syllables chee-vay-loó-a repeated from three to six times in rapid succession. The tone is musical, half sad, half persuasive, beginning somewhat cheerful, and ending more coaxingly.

"Many a night have I spent in the lonely forest and almost uncon-
sciously bowed my head when the chivelua began to sing. A peculiar

Fig. 407. — California quail (*Lophortyx californicus*).

commingling of joy and sadness seems to pervade everything. This is
called by the natives 'la oracion,' or evening prayers, and is one of those
peculiar instances in nature in which the simultaneousness with which
many things take place is wrought into a superstition by man, and by him
regarded as a divine ordinance.

"When the chivelua sings, the golden turkeys fly up in the trees to
roost, the buzzards seek their distant caves, the quails whistle, and all the
birds sing, the bells in the churches toll, and for a few moments all nature
resounds with sweetest music. This lasts during the short tropical twilight

only. The chivelua begins the song; and when he ceases to sing, all song ceases, and not a twitter of bird nor sound of any kind is heard."

The helmet-quails, two species of which are figured here, are beautiful birds from the western and southwestern states and territories. The Californian form extends along the Pacific coast from Washington Territory south to the Colorado River, where it is replaced by an allied species which has the book name Gambel's partridge. Some of the more prominent differences between the two may be seen in the ornamentation of the head, as shown in our next cut. In appearance and habits they are much alike; both have the same erectile plumes on the top of the head; and the following abstract of Dr. Coues's account of Gambel's partridge will do about as well for the other.

Gambel's quail occurs in Arizona and New Mexico, and in some localities is so abundant that one can hardly see how more could find food and cover. Where they

FIG. 408. — Head of Gambel's partridge (*Lophortyx gambeli*).

abound they may be found in almost every sort of a situation, — along the roads, on the sand-heaps, or in the security of the settler's cabin; only the thick pine woods seem unsuited to them. They like the low brush along the creeks; and the mesquite and mimosa scrub has especial attractions for them. Temperature has little effect; for they live both in the burning sands of the desert as well as in the mountains where the snow lies on the ground the year round. For food it is not at a loss; seeds, fruit, and insects all enter into its diet, while the prickly pear and the buds of the willow and mesquite are eaten in their season.

" A beautiful sight it is to see the enamored pressing suit with all the pomp and circumstance of their brilliant courtships. The firm and stately tread, with body erect and comely shape displayed to the best advantage; the quivering wings; the motion of the plumes that wave like the standard of knights-errant; the flashing eyes; bespeak proud consciousness of masculine vigor. The beautiful bird glances defiance, and challenges loudly, eager for a rival; but none disputes, and he may retire, his rights proven. Only a gentler bird is near, hidden in a leafy bower, whence she watches, admiring his bearing, fascinated by the courage she sees displayed, hoping every moment that the next will bring him, dreading lest it may."

If difficulty in shooting increases a bird's gaminess, then Gambel's quail deserves to rank high among the game-birds. It does not flush so readily as our own eastern quail and partridge, and its strong and even flight calls for the highest skill of the gunner. Indeed, were the locality different Gambel's quail would rank high; for, says Dr. Coues, "Here is plenty at least, if not peace. Nothing mars the pleasures of the chase, but the chances of being chased. Were it not for the Indians, we should have here the acme of quail-shooting."

There are many other quail in the west, varying in size and importance. Largest and handsomest is the plumed quail of California and Oregon, with its two long, plume-like feathers curving gracefully over the back. Then there is the blue quail of Arizona, in which the sexes differ but little in appearance, and which has decidedly terrestrial habits, and can run very swiftly, but is not easily flushed. The Massena quail, too, has about the same range, and is remarkable for its utter fearlessness.

What constitutes a quail, and what a partridge, is a question about which there is considerable difference of opinion. Naturalists have reiterated the statement that we have no true partridge in this country, but in spite of all their protests the term continues to live.

FIG. 409.—Plumed quail (*Oreortyx pictus*).

When our ancestors came to this country they applied to its native birds and mammals the terms that they used for somewhat similar forms in their old home, and then were the terms, robin, partridge, and the like, fastened on species which bear but a superficial resemblance to those of Europe. The confusion which this introduced went even farther. Thus the Pilgrim Fathers called one bird a partridge and another a quail, while the ancestors of the first families of Virginia applied the name partridge to the quail of the north. This confusion persists until the present time, and the only escape seems to be to use the name bob-white for the bird known to science as *Ortyx virginianus* — the partridge of Virginia, the quail of New England.

The bob-white is one of the best of our game-birds. It furnishes fine sport for the hunter, and unlike its relatives of the west, seems to know exactly how it should act in the presence of a dog. And then when cooked it has its value: for its flesh is delicious, and can hardly be excelled. It is not a timorous bird; but rears its progeny even in the vicinity of our large cities, and occurs in goodly numbers within five or six miles of

Boston. Anywhere, where it is not hunted to death, it will live year after year regardless of the fact that man is a near neighbor.

Its nest is a rude affair, merely a few sticks and leaves huddled together, and perhaps lined by some bits of bark; and in this it rears a family sometimes of fifteen. Beautiful are the little chicks with their downy coat of brown, and active are they too. They follow their mother about like chickens after a hen, and then when she sounds the note of

FIG. 410. — Bob-white (Ortyx virginianus).

alarm, how instantly they disappear! The mother at such a time has a trick worth knowing. Even though you have been deceived by her a hundred times, this time you think there can be no mistake; she surely has a broken leg or an injured wing. Certainly no sound bird could counterfeit such agony, or give utterance to such notes of pain. She leads you on in this way, and then suddenly her whole aspect changes, and with a whirr she is off, and you realize that you have been sold exactly as you have been so often before. You turn now to look for the chicks; but not one can you see. Their colors are exactly like the dead leaves that strew the ground; but shortly you hear the mother calling at a distance, and you may rest assured that her covey will soon be with her.

Every sportsman will tell you that the quail has the power of withholding her scent. There is certainly something unexplained in the matter, for often the dog with the strongest nose is utterly unaware of the proximity of quail. It is, however, a question whether it really withholds the scent of her own volition, or whether the state of the atmosphere has something to do with it. The question certainly needs farther study before a decision can be reached.

The grouse occur in both hemispheres, but are more numerous with us than they are in the Old World. First in order come the ptarmigan, which, according to Dr. Coues, may be defined as grouse, which turn white in winter. They also have the legs heavily feathered. But two forms need to be mentioned, — the white-tailed ptarmigan of the Rocky Mountains, as far south as 37° N., and the willow-ptarmigan, which lives in the boreal regions of both hemispheres, descending occasionally into northern Maine and New York.

This change of color, which is occasioned by molting, is clearly of great value to these birds, as by it they are more readily enabled to conceal themselves from their enemies. In the summer their dark plumage has the same general color as the gray and lichen-covered rocks which they fre-

FIG. 411. — White-tailed ptarmigan (*Lagopus leucurus*), the upper figure representing the summer, the lower the winter plumage.

quent, while the white winter plumage renders it difficult to see one of these birds in the midst of the snow. How close this resemblance is can be seen by a quotation from Mr. Trippe regarding the white-tailed species: "Sometimes, on seeing one alight on a certain spot, and withdrawing my eyes from it a moment, I have been unable to find it again, although I knew the exact spot where it sat, until a movement on the part of the bird betrayed its position." The ptarmigan are very tame, but will defend their nests and young during the breeding season by flying in the face of any intruder. They make their nests — mere depressions in the ground, lined with a few leaves, straws, and feathers — in rocky regions; and there raise their brood of from four to eight young.

The sharp-tail or pin-tail grouse has much the same range as the white-tailed ptarmigan, reaching east into Minnesota and Iowa. It is mottled above and white below, and on either side of the neck is a blue air-sac, which is called into play in making their strange, booming, deep bass note, sounded in the breeding season. When the cry is uttered, the birds, male and female, gather in some favorable spot, and then begins the 'dance.' One author describes it as a regular 'walk around' as ludicrous

FIG. 412.— Willow-ptarmigan (*Lagopus albus*), in summer plumage.

to the disinterested observer as some of the performances on the comic stage. The males strut around with the wings spread, the feathers all erect, the tail straight in the air, the head nearly resting on the ground, and the vocal sacs inflated and displayed to the utmost. And so the affair goes on, with its display, its strutting, and its occasional battles, until all the birds are mated.

When this affair is over, nest-building is in order; and such care as they exhibit to conceal their home is seldom excelled. It is artfully adapted to its surroundings. "With admirable instinct she will avoid a place that offers such chances of concealment as to invite curious search; her willow-bush is the duplicate of a thousand others at hand; her tuft of grass on the prairie is the counterpart of a million others around; her nest will be found by accident oftener than by design. And when stooping over a warm nest on the prairie, whence she has just fluttered in dismay, we note how exposed it seems now that it is found, we wonder how the dozen blades of grass that overarch the eggs, or the rank weed that shadows them, could have hidden the home so effectually that we nearly trod upon the bird before we saw her." The dingy, yellow-mottled, downy young run as soon as hatched. They rapidly grow, and with winter have attained the size and plumage of the adult.

The name prairie-chicken is frequently given to the species just mentioned; by rights it belongs to another form shown in our cut, and which figures in books and nowhere else as pinnated grouse. It has the same vocal sacs on the sides of the neck, but is able to conceal them by means of small bunches of feathers arising just above them from the sides of the neck. In years gone by this bird extended clear to the Atlantic; but now, with the exception

Fig. 413.—Prairie-chicken (*Cupidonia cupido*).

of a small colony on Martha's Vineyard and possibly on Long Island, they are confined to the west, but stopping ere it reaches Colorado or Wyoming. The amount of slaughter to which it has been subjected is perfectly enormous. It has been hunted by sportsmen, and by others, without the slightest particle of the instinct of the sportsman. It has been netted and trapped; and in the winter the frozen bodies have been shipped east by the car-load. Is it any wonder that it is rapidly disappearing? Unless more stringent laws are passed and, more essential, unless public sentiment enforces these laws, the prairie-chickens will soon be exterminated in the United States.

Another of the grouse is the sage-cock, or cock of the plains, whose home is the treeless desert region of the western territories, where the prominent feature of the landscape is the ever-present wild sage, or Artemisia. It is, excepting the turkey, our largest member of the group of Rasores; but it differs from all its relatives in one point of its structure. All the rest have a strong muscular gizzard familiar to all in the common fowl; but in the sage-cock, as was first pointed out by Mr. Ridgway, it "is soft and membranous like that of the birds of prey. This was first told me by hunters in Nevada; and I afterwards satisfied myself of the truth of their statement, that the sage-hen 'has no gizzard,' by dissecting a sufficient number of individuals. This bird is never known to eat grain; but it subsists almost entirely upon green leaves of Artemisia and on grasshoppers."

The wild sage, as is well known, is bitter, and the diet of its leaves sometimes gives the flesh of the sage-cock a very unpleasant taste. It is said that if one draws the bird as soon as it is killed, this taste will be avoided. The flesh is dark, and rather dry. The vocal sacs, or 'drums,' of this species are perfectly enormous; they are yellow, and when fully inflated have not the smooth half-orange appearance of those of the prairie-chicken, but are large and bulging, with irregular surface, and almost seem to meet beneath the neck.

The ruffed grouse is the 'partridge' of the north, the 'pheasant' of the south, a peculiarity of nomenclature already referred to. It is a rather large bird of brownish plumage, familiar to all. One of the most noticeable features connected with this bird is its habit of 'drumming,' not only at the breeding season, but at other times of the year. Every one has heard the loud sound produced by the bird, and almost every one has seen him in the act. The cock mounts on some log, struts about, and then begins to beat the air with his

Fig. 414. — Ruffed grouse (*Bonasa umbellus*).

wings; slowly at first, but soon the motion becomes more rapid, until the position of the wings is represented by a mere haze, like that produced by

the spokes of a rapidly revolving wheel.   Then the noise is produced, but
the exact way is still a mystery.   Some say the wings hit against the log;
some, against the body; some, that they strike each other over the back;
and others, that the simple beating of the air produces the note.   The
reader may take his choice, or, better, strive to solve the problem for
himself.

Far more important than the grouse and quail is the great group of
pheasants and pheasant-like forms.   None of them are American, but
many are so interesting that we must devote a little space to them.

First come the pea-fowl, natives of India and the Malay regions, one
of the two species of which is familiar to all.   These large birds, with
their bright metallic plumage, are among the most beautiful of all the
feathered tribe.   Of their appearance, habits, and not exactly musical
voice, in the state of domestication, nothing needs to be said.   In India
they live in the jungle, and every hunter confirms the statement that where
these birds, with their resplendent trains are seen, the tiger may be confidently looked for.   What is the association between the two, no one has
yet ascertained.   Beautiful as is our peacock, it is excelled by its Javanese cousin, which has not been introduced into our western world.   The
neck is covered with feathers of a 'scarlet-like green'; the crest is different in shape, while the train is fully as large and fully as beautiful.
The pea-fowl go in large flocks, but their near relations, the more soberly
colored but still beautiful argus pheasants, are more solitary, usually
going in pairs.   One curious fact in the distribution of these forms is
peculiar.   Their range is about the same, but in a region where one is
found the other is a minus quantity.   The following account of their
habits is from the pages of Mr. Wallace: —

The argus makes "a large circus, some ten or twelve feet in diameter,
in the forest, which it clears of every leaf and twig and branch, till the
ground is perfectly swept and garnished,   On the margin of this circus
there is invariably a projecting branch, or high-arched root, at a few feet
elevation above the ground, on which the female bird takes its place, while
in the ring the male — the male birds alone possess great decoration —
shows off all its magnificence for the gratification and pleasure of its
consort, and to exalt himself in her eyes.   It is a strange fact that when
the male bird has been caught, — these birds are much trapped by the
natives [of Sumatra], their excessive shyness making it almost impossible
to shoot them, — the female invariably returns to the same circus with a
new mate, even if two or three times in succession her lord should be
caught.   The female bird is rarely caught, owing to her flying to her
roost when approaching the circus, while the great-winged males walk

ARGUS PHEASANT (*Argus giganteus*).

into the ring, which the native skilfully barricades, except the one spot where he sets his snare."

In treating of the true pheasants words utterly fail. No pen can convey the slightest idea of their unequalled beauty, no brush can reproduce the brilliancy of their plumage. The brightest and most vivid of reds and greens, the most polished silver and gold, is lavishly spread over their bodies. Of the plainer forms, the well-known English pheasant is best known. This is not a native of England, but was brought by the Greeks from Colchis ages ago, and it is probable that it was introduced into the British Isles by the Roman conquerors. There is a record of it there in the year 1059. Of the monals almost nothing is known, except the birds themselves. They have about the same range as the tragopans, but are even more beautiful than they.

Of another style of beauty are the silver pheasants, forest birds with white or silvery backs; the firebacks in which, as the name indicates, the back is of a brilliant fiery red; and the wonderful golden pheasants and their near relatives, the Lady Amherst pheasant, which our cut displays as well as black and white can do it. Its peculiar ruff is white, banded with deep green, the rúmp a golden yellow, its breast a bright, iridescent, metallic green, and the extremely elongate tail-feathers banded and mottled gray, green, and black. But little is known of their habits. They dwell on the mountains and in the deep forests, where the white man rarely penetrates, and are peculiarly hardy. They usually go in small flocks.

Less beautiful, but still far from ordinary birds, are the jungle-fowl of southeastern Asia. The most brilliantly colored and elaborately plumaged of our domestic chanticleers resembles, but does not excel, the jungle-fowl. The resemblance is a natural one; for there now exists but little doubt that our familiar barnyard fowl is one of these jungle-fowl modified by a long life of domestication. From them, too, they derive their pugnacity; for in the jungle they fight as hard as do the best of the game-cocks of so-called civilization. Haeckel describes the jungle-fowl of Ceylon as follows: —

"The melancholy cries of some birds, particularly the green wood-pigeons and bee-eaters, are rarely heard except at early dawn; at a later hour the brilliant jungle-fowl is the only bird that breaks the silence. This gorgeous species appears to be nearly related to the first parent of our domestic fowl. The cock is conspicuous for his gaudy and brilliant plumage; fine orange-brown ruff, and green sickle-shaped tail-feathers; while the hen is bedecked in a modest grayish brown. The sonorous call of this wild fowl, fuller and more tuneful than the crow of his barnyard cousin, is frequently heard in the woods for hours, now near, now far away; for the rival cocks compete in this vocal concert for the favor of the critical hens. I could, however, but rarely get within gunshot of them; for they are so shy and so cautious that the slightest rustle disturbs them and stops the concert."

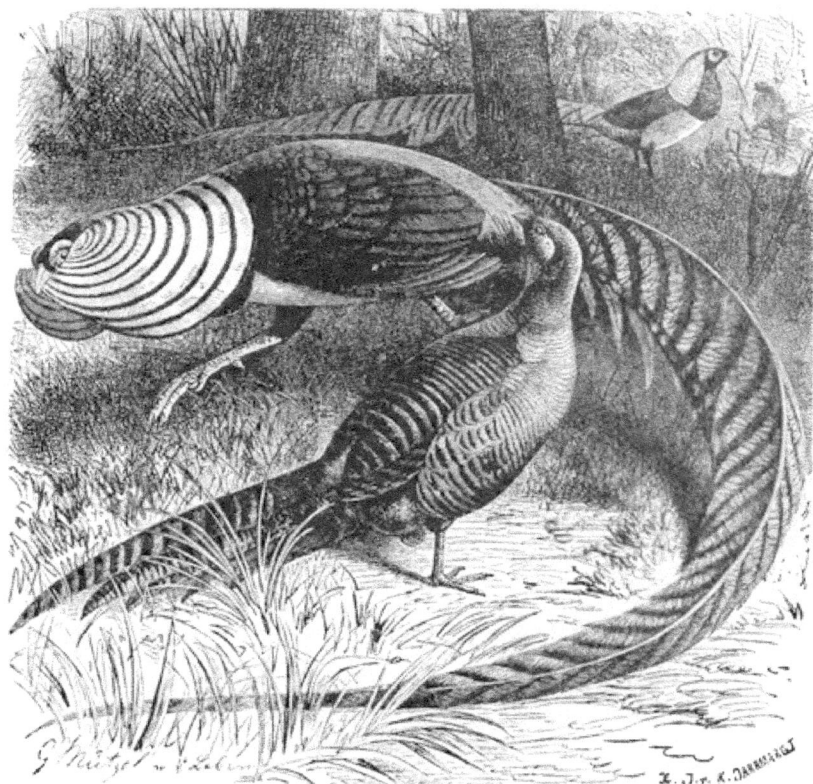

FIG. 415. — Lady Amherst pheasant (*Thaumalea amherstiæ*).

The prettily mottled, but harsh-voiced guinea-fowl are all natives of Africa, two species reaching over into Madagascar. Concerning their habits there is but little of interest to be said.

Not so with the turkeys, the three species of which are American. Every believer in Thanksgiving Day and Christmas has an interest in them; for they, and not the eagle, should be our national bird. Our domesticated turkey is much unlike the wild bird from which it has descended, and which once roamed over the whole United States. It has changed in form and color, and yet this change takes place in a very few generations, and an even shorter time is necessary for escaped birds to revert to their former condition. In their wild state they are very timorous; at night they roost on the tallest trees, and the hen, on leaving her nest, carefully covers it with leaves. Our native bird, with its plumage of dark bronze-green, is handsome, but he is far excelled in this respect by the species of Mexico and Central America, the latter being especially brilliant, and even rivalling the pheasants in its plumage.

FIG. 416. — Central American turkey (*Meleagris ocellata*).

In all the scratching birds so far enumerated the hind toe is placed a little above its fellows, but in some birds of South America and the far east it is on the same level with the rest, as is the case in the doves. There are one or two of these species so interesting in their habits that they must not be passed by. The first are the megapodes, or brush-turkeys, the first name having reference to the size of their feet.

The megapodes are found in Australia, and north into the Philippines and Borneo. They are all characterized by large feet and long, curved claws, which most of the species put to a most peculiar use. As described by Mr. Wallace, they "rake and scratch together all sorts of rubbish, dead leaves, sticks, stones, earth, rotten wood, etc., till they form a large mound, often six feet high and twelve feet across, in the middle of which they bury their eggs. The natives can tell by the condition of these mounds whether they contain eggs or not; and they rob them whenever they can, as the brick-red eggs (as large as those of a swan) are considered a great delicacy. A number of birds are said to join in making these mounds, and lay their eggs together, so that sometimes forty or fifty may be found. The mounds are to be met with here and there in dense thickets, and are great puzzles to strangers, who cannot understand who can possibly have heaped together cart-loads of rubbish in such out-of-the-way places; and when they inquire of the natives they are but little wiser, for it almost always appears to them the wildest romance to be told that it is all done by birds."

The maleo, a native of Celebes, has a different habit in laying its eggs. In certain parts of the island there are large beds of volcanic sand, which, under the vertical sun, becomes very hot. To these sandbanks the birds repair in August and September, coming from a distance of ten or fifteen miles to find a suitable spot. Here both cock and hen unite in digging a hole for the eggs, and then cover them up with about a foot of sand. One egg is laid at a time, and at the end of ten or twelve days the hen returns to lay another, and so on until she has laid her complement. Here the eggs are left to be hatched by the heat, and the young birds, on escaping from the shell, quickly work their way to the surface, and immediately disappear in the forest. They can even fly as soon as hatched.

Concerning the origin of this habit, so contrary to that of most related birds which incubate their eggs, Mr. Wallace has some very interesting observations. "Each egg," says he, "being so large as to entirely fill up the abdominal cavity, and with difficulty pass the walls of the pelvis, a considerable interval is required before the successive eggs can be matured. . . . Each bird lays six or eight eggs, or even more, each season, so that between the first and last there may be an interval of two or three months. Now if these eggs were hatched in the ordinary way, either the parent must continue setting for this long period, or, if they only began to set after the last egg was deposited, the first would be exposed to injury by the climate, or to destruction by the large lizards, snakes, or other animals which abound in the district, because such large birds must roam about a good deal in search of food. Here, then, we seem to have a case in which the habits of a bird may be directly traced to its exceptional organization; for it will hardly be maintained that this abnormal structure and peculiar food (fallen fruits) were given to them in order that they might not exhibit that parental affection, or possess those domestic instincts so general in the class of birds, and which so much excite our admiration."

The currasows and their relatives all belong to tropical America, only one, the Texan guan, entering the United States. All of the group are essentially forest birds; and most of them live habitually in the trees. The true currasows are large and handsome birds of the tropics. They live in the tops of the highest trees, where they search for their food. Only one is an exception to this, as it lives during the day in burrows beneath the ground, and at night it comes out and joins the others in the top of the forest. In reference to these forms, two very different statements appear. One author says that the natives of the country in which they abound frequently tame these birds, and that they breed well in captivity. Another says that they are occasionally tamed, but not domesticated, as they do not breed in confinement. The flesh is rather dry and not especially well flavored.

SAND-GROUSE.

The sand-grouse are merely mentioned to let the reader know that such a group exists. They are inhabitants of tropical Africa and Asia, and in their appearance are somewhat intermediate between a grouse and a pigeon; but they have such a peculiar structure that naturalists have placed them in a group by themselves. They fly in large flocks, but are not especially good for food.

PIGEONS.

Before taking up the typical pigeons, we must consider one of the least pigeon-like forms, — the celebrated dodo. Two centuries and a half ago this huge, ungainly bird was abundant on the Mascarene Islands. — Bourbon, Mauritius, and Rodriguez; but before 1700 it was extinct. We know something of how the bird looked, both from pictures and descriptions. Apparently several were brought to Europe; and one certainly was exhibited in London in 1638. There was, too, a stuffed specimen in the museum of Oxford University; but during the present century some of the governing board thought its condition disgraceful, and they ordered it destroyed. The head and legs alone escaped; and this, together with a number of bones from a swamp in Mauritius, and a head in Copenhagen, are all that is left. The introduction of dogs and hogs into the Mascarenes wrought the destruction of this flightless bird.

The quaint account of Bontius is an example of many: "The Dronte or Dodaers is for bigness of mean size between an Ostrich and a Turkey, from which it partly differs in shape and partly agrees with them, especially with the African Ostriches, if you consider the rump, quill, and feathers; so that it was like a pigmy among them if you regard the shortness of its legs. It hath a great ill-favoured head, covered with a membrane resembling a hood; great black eyes; a bending prominent fat neck, an extraordinary long, strong, bluish white bill, only the ends of each mandible are of a different colour, that of the upper black, that of the nether yellowish, both sharp-pointed and crooked. Its gape, huge wide, as being naturally very voracious. Its body is fat and round, covered with soft gray feathers after the manner of an ostrich; in each side, instead of hard wing-feathers or quills, it is furnished with small soft-

Fig. 417.—Tooth-bill pigeon (*Didunculus strigirostris*).

feathered wings of a yellowish ash colour; and behind the rump instead of a tail is adorned with five small curled feathers of the same colour. It hath yellow legs, thick but very short; four toes in each foot, solid, long, as it were scaly, armed with strong black claws. It is a slow-paced and stupid bird, which easily becomes a prey to the fowlers. The flesh, especially of the breast, is fat, esculent, and so copious that three or four dodos will sometimes suffice to fill one hundred seamen's bellies. If they be old or not well boiled, they be difficult of concoction, and are salted and stored up for provision of victual." Another author says that " meat it is with some, but better to the eye than stomach, such as only a strong appetite can vanish."

Every little while the story is told that some voyager to the South Seas has discovered the dodo alive, — the last repetition was in 1886, — but in every instance it turns out that the bird was considerably different, came from a different locality, and was well known before as the tooth-billed pigeon of the Samoan Islands. It is a smaller bird than the dodo, a strong flyer, and has a plumage of a reddish brown. An especially interesting fact in connection with it is its recent change of habits. When first known it was a terrestrial bird, spending most of its time on the

FIG. 418. — Blue-headed pigeon (*Starnænas cyanocephala*).

ground; but since cats and rats were introduced, it lives almost exclusively in trees; and hence instead of rapidly approaching extinction it is becoming more abundant.

The blue-headed pigeon, or blue-headed quail-dove, is a beautiful species, living in Cuba; but occasionally crossing to Florida on the one hand, and Jamaica and the other West India Islands on the other. Its general color is a reddish brown, the patch on the breast black, the line beneath the eye white, and the top of the head blue.

As a rule, ground-feeding birds are poor flyers; but the beautiful Nicobar pigeon of the eastern seas is an exception. It is a copper and green species, with a white tail, and beautiful pendant feathers on the neck. It looks so heavy that one would hardly think it could fly a mile, and yet it is really very strong on the wing; and one was known to have flown from New Guinea to a small island one hundred miles to the north, with no intervening land.

The mountain-witch, says Mr. Gosse, is the most beautiful bird, the long-tailed humming-bird excepted, in the island of Jamaica. Gray, brown, buff, bluish gray, pale crimson, brassy green, dark red, the richest purple, deep sea-green, black, bistre-chestnut, are all represented in its plumage. It is of a retiring disposition, flying but poorly, and making its home in the deepest woods on the mountains. "Its coo consists of two loud notes, the first short and sharp, the second protracted and descending with a mournful cadence. At a distance its first note is inaudible, and the second reiterated at measured intervals sounds like the groaning of a dying man. These moans, heard in the most recluse and solemn glens, while the bird is rarely seen, have probably given it the name of mountain-witch." A closely related bird of the same island is the partridge-dove, which like the mountain-witch eats the seeds of the castor-bean, and the physic-nut among other things.

A third species from Jamaica is the white-belly. It lives in the higher parts of the island, "where its loud and plaintive cooing makes the woods resound. The negroes delight to ascribe imaginary words to the voices of birds, and indeed for the cooings of many of the pigeons this requires no great stretch of the imagination. The beautiful white-belly complains all day, in the sunshine as well as the storm, 'Rain-come-wet-me-through!' each syllable uttered with a sobbing separateness, and the last prolonged with such a melancholy fall, as if the poor bird were in the extremity of suffering." It is a very gentle species, even when first caught; its food is the physic-nut, and the seeds of the orange and mango. The pea-dove says 'Sary-coat-true-blue'; while the white-wing has a still larger vocabulary. At times it says 'two bits for two,' or 'what's that to you'; but at other times the sentence runs (according to negro interpretation), 'Since poor Gilpice die, cow-head spoil,' the word 'spoil' prolonged and falling in tone, as if the spoiling of the cow-head were a most deplorable circumstance.

There are some twenty-five or thirty species of turtle-doves in the Old World, one of which has attained considerable celebrity. It is a meek species, utterly lacking in spirit, and displaying the greatest affection for its mate. While many of the pigeons display some spirit, this one has been universally recognized as the personification of peace.

The Carolina, or common dove of the United States, is far better known to us, distributed as it is over nearly the whole of the United States. Its home is in the south, but each spring it migrates to the north, and some even pass the winter in New England. The flocks arrive at their summer home about the last of April or the first of May, and remain until October. It is frequently called the mourning-dove from the melancholy character of its cooing notes.

Even more celebrated is the wild or passenger pigeon, which formerly ranged over our country in vast flocks, but which has now become greatly reduced in numbers, although it still abounds in the more newly settled localities. For a picture of this bird as it used to be, we must refer to the older writers on ornithology, and of these none has given a more graphic account than the father of American bird-lore.

Fig. 419. — Passenger-pigeon (Ectopistes migratoria).

Alexander Wilson. They are remarkable for their migratory habits, and their associating together in flocks so large that no man can begin to enumerate them. Single flocks have been seen which must have contained hundreds of millions! Speaking of what he had seen in Ohio, Kentucky, and Indiana, Wilson says: "These fertile and extensive regions abound with the nutritious beech-nut, which constitutes the chief food of the wild pigeon. In seasons when these nuts are abundant, corresponding numbers of wild pigeons may be confidently expected. It sometimes happens that, having consumed the whole product of the beech-trees in an extensive district, they discover another at a distance of perhaps sixty or eighty miles, to which they regularly repair every morning, and return as regularly in the course of the day, or in the evening, to the general place of rendezvous, or, as it is usually called, the roosting-place. These roosting-places are always in the woods, and sometimes occupy a large extent of forest. When they have frequented one of these places for some time, the appearance it exhibits is surprising. The ground is covered to the depth of several inches with their dung; all the tender grass and underwood destroyed; the surface strewed with large limbs of trees broken down by the weight of the birds clustering one above another; and the trees themselves, for thousands of acres, killed as completely as if girdled with an axe."

One of their breeding-places not far from Shelbyville, Ky., " stretched through the woods in nearly a north and south direction, was several miles in breadth, and was said to be upwards of forty miles in extent! In this tract almost every tree was furnished with nests wherever the branches could accommodate them. . . .

"As soon as the young were fully grown, and before they left their nests, numerous parties of the inhabitants, from all parts of the adjacent country, came with wagons, axes, beds, cooking-utensils, many of them accompanied by the greater part of their families, and encamped for several days at this immense nursery. Several of them informed me that the noise in the woods was so great as to terrify their horses, and that it was difficult for one person to hear another speak without bawling in his ear. . . . The view through the woods presented a perpetual tumult of crowding and fluttering multitudes of pigeons, their wings roaring like thunder, mingled with the frequent crash of falling timber; for now the axe-men were at work cutting down those trees that seemed to be most crowded with nests, and contrived to fell them in such a manner, that in their descent they might bring down several others, by means of which the falling of one large tree sometimes produced two hundred squabs, little inferior in size to the old ones, and almost one mass of fat."

One more quotation must suffice, interesting as the subject is. Wilson saw an immense flock in northern Kentucky, which he describes as follows: " From right to left as far as the eye could reach the breadth of this vast procession extended; seeming everywhere equally crowded. Curious to determine how long this appearance would continue, I took out my watch to note the time, and sat down to observe them. It was then half-past one. I sat for more than an hour, but instead of a diminution of this prodigious procession, it seemed to increase both in numbers and rapidity; and, anxious to reach Frankfort before night, I arose and went on. About four o'clock in the afternoon I crossed the Kentucky River at the town of Frankfort, at which time the living torrent above my head seemed as numerous and extensive as ever. . . . If we suppose this column to have been a mile in breadth (and I believe it to have been much more), and that it moved at the rate of one mile in a minute; four hours, the time it continued passing, would make its whole length two hundred and forty miles. Again, supposing that each square yard of this moving body comprehended three pigeons; the square yards in the whole space, multiplied by three, would give two thousand two hundred and thirty millions, two hundred and seventy-two thousand pigeons. An almost inconceivable multitude, and yet probably far below the actual amount."

The days of such enormous flocks have gone by, and to man is to be attributed the change. Not that the amount that he slaughtered, or the young that he killed, had any appreciable effect; the result was accomplished in a more indirect manner. Man cut off the forests of beech, and as the supply of mast was thus diminished, the number of the pigeons was correspondingly reduced.

Fig. 420. — Rock-pigeon (*Columba livida*).

One of the most remarkable facts in the whole realm of natural history is the remarkable variation to be seen in the domesticated pigeons, and yet all are descended from the rock-pigeon shown in our cut. I cannot conceive how any one can appreciate even this one fact and not believe in evolution. Here from one form species and genera innumerable have been formed in a comparatively limited space of time. The 'nuns,' with their heavily feathered feet, the 'pouters,' with their enormous inflated breasts, the 'fantails,' with far more than the ordinary number of tail-feathers, and a hundred more, would each be made a distinct species or genus, or even a group of higher rank, if found in the state of nature. Why should they not when produced by the action of man?

BIRDS OF PREY.

Once the birds of prey were exalted to the head of the whole series of birds; now they are degraded to about the middle of the avian hosts. Yet whether they rank high or low, they still are interesting, and one, by some

Fig. 421.— Secretary-bird (*Gypogeranus serpentarius*).

great mistake, has been selected as our national emblem. The group — called Accipitres, or Raptores — is well marked, and no one, seeing one of these birds, can in the least mistake its character. The large, strong-hooked beak, the strong toes, the whole physiognomy, at once proclaims the bird of prey.

First in the series is a curious bird from South Africa; long-legged, long-necked, and long-tailed, and with a bundle of long feathers on the back of the head, which, when erected, as they are when the bird is excited, bear a slight resemblance to a number of quill-pens stuck behind the ear, thus giving the bird a clerical appearance. From this they derive their name of secretary-bird.

This bird is very valuable to the inhabitants near the Cape, and is frequently domesticated by them on account of its ridding their premises of

Fig. 422.—Carrion-crow, or black vulture (*Cathartes atrotus*).

vermin of all sorts. Frogs, toads, and lizards are favorite articles of diet; no sort of insects is despised, but esteemed as highly as anything else are snakes. Even the most venomous of these are killed. In attacking them the secretary walks slowly up, and when the reptile strikes, the blow is escaped by an artful dodging, or is parried by the long quills of the wings, and since these have no blood-vessels, no bite can poison the bird.

Next in order come the American buzzards, and concerning these seemingly disgusting creatures we can find many points of interest. But first we must mention a few of the species, so that we may more intelligently consider some of the habits common to all. First comes the carrion-crow, or black vulture, of the south. The carrion-crow is an inhabitant of tropical and sub-tropical America, ranging north to the Carolinas, and occasionally straying even into New England and as far as Maine. The warmer climates, where everything decays so rapidly, form its proper home, and there, like its relatives, it does an immense amount of good as a scavenger. In some places this office of the bird is recognized by legal enactment, and statutes are in force which prohibit its destruction. As its name indicates, it is a black bird; its wings have a spread of nearly five feet, and its neck and

Fig. 425.—Turkey-buzzard (*Cathartes aura*).

head, like those of the other vultures, are nearly bare of feathers. In the southern states they sometimes congregate in large numbers, especially when some attractive carcass lies displayed before them. On one occasion nearly three hundred were counted, perched upon every available object, waiting for a chance to attack the body of a freshly flayed horse.

A somewhat larger bird, and one with a browner plumage, is the turkey-buzzard, which has much the same range; that of the present bird extending much farther to the west. Where the carrion is they gather like their cousins. At other times they may be seen perched upon some tree, or circling with that apparently magical motion high in the air. Round and round they go, far above the earth, their wings never beating as do those of the black vulture ; and all the time they are keeping a sharp lookout for the good things of earth. Such an eye as they have ! When the bird is near at hand, one notices nothing strange in it : but suspended half a mile above the earth, nothing escapes it. The smallest striped snake winding its way through the grass, or perhaps across the beaten road, is quickly seen.

Still larger is the California vulture, one of the largest birds of this continent. It is much like its cousins of the east.

In tropical America, are two species which must not be omitted : these are the king-vulture, which lives in the forests of South America ; and the still more celebrated condor of the Andes, about which many wonderful stories — stories more wonderful than true — are told. This latter species is often said to be the largest bird in existence, and some of the tales of its size would make it a rival of the roc of Marco Polo, if not of the bird of the same name which plays such an important part in the pages of the 'Thousand and One Nights.' So, too, it is said to carry off children, sheep, and the like to its nest in the highest peaks of the mountains. Look for a moment at our picture of the condor, and consider the character of its feet. The claws are short and nearly straight, and have nothing of the character of the curved talons of those birds of prey which are accustomed to carry heavy weights in their feet. Then, next listen to the facts about the size of the birds. With the exception of a doubtful statement by von Tschudi, no condor when brought to that surest test of size, the measuring-tape, has been found to exceed twelve feet in expanse of wing ; and even von Tschudi's specimen spread only fourteen feet and ten inches. Does it now seem probable that a bird so small as this could carry off a sheep or a child ?

These stories, however, remain to be explained. First as to size. Doubtless the transparency of the air, and the absence of any known objects for the purpose of comparison in the treeless regions, have led to exaggeration in this respect, and Humboldt found that birds which actually measured less than four feet from the tip of the beak to the end of the tail, appeared of enormous size when seen at a distance perched upon some rock. The stories of their rapacity are to be explained in only one way. Some traveler who wished to make a good story, took 'whole cloth' for the purpose.

All of these vultures are carrion-feeding birds ; but, contrary to the oft-repeated statement that they feed on this alone, they are fond of fresh meat as well ; and they will frequently kill small animals for the sake of food. The turkey-buzzard, by the way, is fond of the eggs of the alligator, and devours them whenever found. To the statement that they wait for a carcass to decompose before attacking it, and hence must prefer tainted meat to fresh, the reply is that they cannot tear the sound hide, but must wait until decomposition makes an opening for them.

FIG. 424. — Condor (*Sarcorhamphus gryphus*).

Then as to the sense of smell. It was formerly affirmed that their olfactory powers were unsurpassed. Their nostrils are large, and their olfactory nerves are highly developed; and yet it would appear that sight is far more important in guiding them to food than is the sense of smell. The other view, however, prevailed until 1826, when Audubon laid before a scientific society in Edinburgh his experiments in this direction. The theory that carrion-feeding birds were attracted by the strong odor of the food was a good one — until the subject was tested.   Some of Audubon's experiments were subsequently repeated by his fellow-laborer, the Rev. John Buchman, who also added others.

To show that sight was the sense depended upon, a stuffed deer, dry and without a particle of odor about it, was placed in the position of death, and only a few moments elapsed before a vulture was on hand, tugging and pulling at the hide until he became disgusted.   Again, a hastily painted carcass, representing the body of a sheep deprived of its skin and with its body cut open, was carried out into the open air, and proved a great attraction to these birds; and they tried their best to get at it, and to tear it in pieces with their bills and claws.

The experiments to show that smell was not the guide were equally conclusive.   The carcass of a hog, in a fairly putrid condition was concealed under a pile of brush; and although the vultures flew quite near, they never discovered it.   Again, a wheelbarrow-load of carrion, covered with a sheet, was taken to the fields; and upon the sheet bits of fresh meat were strewn.   These the birds soon made way with; but although their bills were often separated from the carrion beneath the sheet by but the thickness of the cloth, they utterly failed to find it, and showed no unmistakable signs that they recognized its presence anywhere in the neighborhood.

There is a little evidence on the other side which must not be neglected. Best is the testimony of Mr. Richard Hill of Jamaica.   In this case a poor German, living by himself, fell sick with a fever after he had placed meat and vegetables in a pot for a stew.   For two days he lay senseless; and the meat putrefied.   "Vulture after vulture as they sailed past were observed always to descend to the cottage of the German, and to sweep round, as if they had tracked some putrid carcass, but failed to find exactly where it was."   Again, Waterton in South America killed lizards and frogs and put them in the way of the buzzards; but they did not appear to notice them until they became putrid.

Still, if they are guided by scent, why is it that they are not continually chasing each other; for all agree that the smell of carrion is not much worse than the odor of these birds, which Pennant says, "are most unpleasantly foetid."

There are vultures in the Old World, but they are much different from those of America.   In fact, they are merely eagles, modified for a diet of carrion.   They have the same bare head, the same weak claws, as do their representatives with us.   In the warmer parts of their country they act as scavengers, just as the carrion-crow and the turkey-buzzard do with us; and, in short, their habits are so much alike that it would be but mere repetition to detail them here. The eagle group will be much more interesting.

In Europe and Asia the Lammergeyer replaces the condor of the Andes, tough stories and all. It is a large bird, more like a vulture than like an eagle in its habits. It makes its home in the high mountains, — the Alps and Pyrenees in Europe, — building its nest usually on some inaccessible cliff. The bird feeds on anything it can pick up, — carrion refuse of all sorts, snakes, and small mammals; and it is said that it has been known to attack children, but the stories of its carrying babes off to its mountain nest seem to exist only to point a moral or adorn a tale.

The golden eagle is common to all the northern countries of the globe, reaching south with us to about the thirty-fifth parallel of latitude, but

FIG. 425. — Bald eagle (*Haliætus leucocephalus*).

nowhere is it very common. It is a strong flier, but is excelled in this respect by the bald eagle and many of the hawks. It makes its nest in inaccessible places, and its eggs are among the greatest desiderata of the collectors.

The better known is the bald eagle, which, as has often been insisted upon, is not bald at all, but merely has a crown of white feathers. Still the name bald eagle will remain in spite of every protest of the purists and every attempt to foist the name of white-headed eagle upon the bird. This is the species which has been chosen as our national emblem; a most unfortunate choice, if character is to count for anything. It is a cowardly bird; not over-choice in the matter of food, and at least a little apt to be tyrannical, or even piratical when opportunity offers. Listen for a moment to Alexander Wilson, telling you how he obtains the fish of which he is so inordinately fond : —

"In procuring these he displays, in a very singular manner, the genius and energy of his character, which is fierce, contemplative, daring, and tyrannical, — attributes not exerted but on particular occasions, but when put forth, overpowering all opposition. Elevated on the high, dead limb of some gigantic tree that commands a wide view of the neighboring shores and ocean, he seems calmly to contemplate the motions of the various feathered tribes that pursue their busy avocations below, — the snow-white gull slowly winnowing the air; the busy sandpipers coursing along the sands; trains of ducks streaming over the surface; silent and watchful cranes intent and wading; clamorous crows; and all the winged multitudes that subsist by the bounty of this vast liquid magazine of nature. High over all these hovers one whose action instantly arrests his whole attention. By his wide curvature of wing and sudden suspension in air, he knows him to be the fish-hawk, settling over some devoted victim of the deep. His eye kindles at the sight; and, balancing himself with half-opened wings on the branch, he watches the result. Down, rapid as an arrow from heaven, descends the object of his attention, the roar of its wings reaching the ear as it disappears in the deep, making the surges foam around. At this moment the eager looks of the eagle are all ardor; and, levelling his neck for flight, he sees the fish-hawk once more emerge, struggling with his prey, and mounting into the air with screams of exultation. These are the signals for our hero, who, launching into the air, instantly gives chase, and soon gains on the fish-hawk; each exerts his utmost to mount above the other, displaying in these rencontres the most elegant and sublime aërial evolutions. The unencumbered eagle rapidly advances, and is just on the point of reaching his opponent when, with a sudden scream, probably of despair and honest execration, the latter drops his fish. The eagle, poising himself for a moment, as if to take a more certain aim, descends like a whirlwind, snatches it ere it reaches the water, and bears his ill-gotten booty silently away to the woods." This is the bird of freedom!

I never saw the eagle more himself than once with a captive specimen. He was in a large slat cage, standing in the yard, and usually appeared rather stupid, except when tormented by the boys, or excited by his regular meals of fish. One day, however, he showed all his aquiline traits, so far as the surroundings would permit : a stray cat, chased by a dog, fled to his cage for refuge. Ere she entered it his whole nature seemed to change; he was no longer sluggish, but every muscle, every feather, seemed alive; and his harsh scream rasped the ear. As the cat entered, the eagle, yelling its loudest, darted like an arrow down from his perch, and quicker than one can tell it, had his long, curved talons driven clear through the unfortunate pussy, while the strong, hooked beak was almost instantly employed in tearing off the head. The cat had only time to give one screech, and she was dead. Had he been at liberty he could not have displayed more of himself, while, of course, I could have had no such opportunity of observation. When the cat appeared, the cage was forgotten; all that he thought of was game.

Fig. 426. — Guiana eagle (*Morphnus guianensis*).

Two eagles from tropical America need mention. They are provided with a crest of erectile feathers on the top of the head, and when this crown is expanded, the naturally fierce appearance of the bird is greatly increased. The Guiana eagle of the tropical forests of the Amazon and the Orinoco, is the smaller bird of the two. The harpy eagle, which ranges from southern Mexico, and even southern California, south to southern

Brazil, is much larger, but both are much alike in general appearance. The harpy is one of the strongest birds in the world. It will attack almost anything. Full-grown turkey-cocks, sloths, foxes and badgers, pigs and dogs, are game for it; and even the large sapajou monkey, three times the size of the harpy, stands no chance. They may even give a man a hard struggle. Dr. Felix Oswald has given a most excellent description of the habits of these birds, and relates the experience which a Mexican miner had with a couple of them. He knocked one down with his cudgel, and put an end to its struggles with a few well-directed blows. Shoulder-ing his game he started down the mountains; but in the most perilous part of the descent he suddenly felt the claw of the eagle at his neck. He managed to free himself, and grasping the bird by his legs, pounded its head against the stones until its struggles ceased. At last he reached the village and deposited the bird on the ground, when it revived again, struck its claws through the hand of its captor, struggled to its feet, and would have escaped after all had not the now thoroughly enraged miner flung himself upon it, and with a stone hammered its head to a jelly.

The harpy is absolute ruler in his domain. He tolerates none of that interference by other birds. Excepting the iris-crow no other bird pesters him, as so many do the eagles and hawks of our own latitudes. The eagles and the owls must keep out of his way, and the sea-eagle is pursued for miles, whenever he ventures into the region of the harpy. The harpy is impatient of any competition, for competition reduces the supply of food. The adult bird devours an immense amount; but the young are even more exacting. "The callow harpies," says Dr. Oswald, "with their pendant crops, their misshapen, big heads, and their preposterous claws, resemble embryo demons, or infantine chimæras rather than any creatures of nature; but they grow very rapidly, and their appetite during the first six months of their existence is almost insatiable." It kept one Indian boy constantly busy to feed a couple of young birds.

The squirrel-hawk and the rough-legged buzzard are first-cousins. The former is confined to the Pacific slope; but the other ranges over the whole of northern North America, and Europe as well. The rough-legged buzzard is rather nocturnal in its habits, and feeds upon pretty small game, which others of its relatives would despise. It usually neglects the larger birds and mammals, and pays its attention to mice and moles, frogs and snakes.

The term 'hen-hawk' is rather elastic and indefinite, and is usually applied to some two or three species of birds, which when occasion offers swoop down from on high, on the poultry in the barnyard. These species are known by various book names, as the red-tailed hawk, red-shouldered hawk, and the like. In habits all are very similar, and a description of one will answer for all. They sail in broad circles high in the air, and exhibit perhaps as well as any other common birds the phenomenon of soaring to which allusion has already been made. From their position on high they are constantly scanning the ground beneath, looking out for

Fig. 427. — Rough-legged buzzard (*Archibuteo lagopus*).

some chicken, or a far more acceptable meal — a rabbit or a squirrel. They are much bolder, more aggressive, and far more active than the rough-legged buzzard.

The fish-hawk, or osprey, which, with its young, is admirably depicted on our plate, is one of the most interesting of all the birds of prey. It is distributed over almost the whole world, and everywhere its habits are much the same, due allowance being made for the change in environment.

With us they are most abundant along the sea-shore, making their nests
in trees near the water.   They are eminently sociable birds, and frequently
several may be found nesting near each other, and hundreds have been
known to live in one community.

The osprey is a hard worker.   One may see it circling around with
sharp, quick strokes over the waves, watching for its daily food.   Soon it
sees a fish below, and then the sight must be seen to be appreciated.   The
wings are closed tight against the body, and the bird throws himself head-

FIG. 428. — Common European buzzard (*Buteo vulgaris*). Our Swainson's buzzard is probably but a
variety of this, and cannot be distinguished from it in the cut.

long down to the water, turns his body quickly, and grasps his prey with
his talons.   Sometimes he plunges a foot, or even more, below the surface in
his efforts to capture the fish.   Even after he gets his dinner, he is not always
sure of it until it is safely stored away in his stomach.   The bald eagle, as
has already been described, often robs him of his booty, while farther south
the frigate-bird plays the same piratical part.   In all parts of the world it
is the same.   Everywhere there are birds ready to take advantage of the
fish-hawk's industry.   These robberies are highly interesting and exciting to
watch, but invariably one's sympathies are with the osprey.   As the robber

FISH-HAWK, OR OSPREY (*Pandion haliætus*).

dashes in upon the scene, the fish-hawk redoubles its efforts to escape. Quicker and more vigorously the long wings beat; higher and higher the spiral flight extends; but all in vain. The fish-hawk, even when unincumbered, is no match for his pursuer; but weighted down by his fish, even though it be a small one, is too heavily handicapped to continue the unequal contest for a long time. Still he keeps up his shrieks, and when, at last, he drops his dinner and wends his way to the shore, his cry is a mixture of disappointment, rage, and a thousand other emotions.

Captain M'Caulay relates an instance of a fish-hawk deceiving an eagle, which is too good to remain buried in the pages of the 'Geological Survey Bulletin.' A friend, he says, noticed a fish-hawk rise from the water with a prize in his mouth, and, after getting a short distance inland, beset upon by an eagle, evidently waiting for a meal, and a quiet spectator of the scene. Being attacked and compelled to give it up, he dropped it, and the eagle, catching it in the air, flew away with it, apparently disregarding the pangs of a guilty conscience. The next day he noticed a repetition of the fishing operation by the hawk, and on the approach of the eagle, as before, the prey was promptly dropped, while the hawk quietly disappeared. The eagle again caught the object in the air, and quickly let it go again. This being thought somewhat strange, led to an investigation, when it was found that the cause of this peculiar conduct, on the part of the thievish bird of freedom, was that the supposed fish was in reality a piece of dried manure.

The true kites all belong to the Old World, and their soaring flight has given their name to the paper toys which boys are so fond of flying in the early days of spring. This similarity of names — the kite of paper and the kite of flesh and feathers — causes foreigners some perplexity, as is the case with many other similar words in the English language; and, so the story goes, a learned German ornithologist soberly catalogued Franklin's famous experiments with a kite in a bibliography of works relating to birds.

We have some birds which are called kites, and two of the species found in the central and southern portions of our country are represented in our figure. One of these, the swallow-tailed kite, deserves more space than we can afford it; for it is at once the handsomest species and the most graceful flier of all our birds of prey. We cannot forego one short quotation. The swallow-tailed kite " courses through the air with a grace and buoyancy it would be vain to rival. By a stroke of the thin-bladed wings and a lashing of the cleft tail, its flight is swayed to this side or that in a moment, or instantly arrested. Now it swoops with incredible swiftness, seizes without a pause, and bears its struggling captive aloft, feeding from its talons as it flies; now it mounts in airy circles till it is a speck in the blue ether and disappears. All its actions, in wantonness or in the severity of the chase, display the dash of the athletic bird, which, if lacking the brute strength and the brutal ferocity of some, becomes their peer in prowess — like the trained gymnast, whose tight-strung thews, supple joints, and swelling muscles, under marvellous control, enable him to execute feats that to the more massive or not so well conditioned frame would be impossible."

Fig. 429.— Mississippi kite (*Ictinia subcærulea*), left; and swallow-tailed kite (*Elanoides forficatus*), right.

In India is a kite which is a feature of every landscape. It is the omnipresent pariah kite. As a scavenger it does good service, but its ubiquity and its boldness make it something of a nuisance. It robs the people of the food which, in the fashion of the country, they may be carrying home from the market on the head; it will even attempt to seize a morsel from the hand of man, and whether successful or not will be far away before the victim has begun to recover from his surprise. It also takes fishes and frogs, reptiles of all sorts, and small birds and mammals when occasion offers.

A small European kite deserves mention from its peculiar diet. It cares but little for the larger game of its relatives; but if it can obtain insects or insect larvæ, it asks nothing better. It attacks the nests of wasps and bees, utterly regardless of the stings of the infuriated insects, and fills itself with the larvæ which it finds in the comb. This habit has given it the name of bee-kite.

In arranging the birds of prey, naturalists are fond of calling certain birds "true," and others "so-called." Thus we have true and so-called buzzards, true and so-called vultures; and so on through the whole series. Following the time-honored custom, we will speak of a few of the 'true' hawks. Here comes the goshawk, which with its varieties occurs in both Europe and America. It is a large bird, and flies strongly and rapidly, and is not accustomed to soaring, like so many of its relatives. Here also comes Cooper's hawk and the sharp-shinned hawk, and among others two African hawks, which are remarkable among all the birds of prey from their having a truly musical voice, and of singing its short notes at intervals for hours together. None of the true hawks are large; and some are very small. A foot and a half from the tip of the beak to the end of the tail is but rarely exceeded among them, except in the case of the goshawk, the female of which may be two feet long.

Fig. 430. — Young goshawk (*Astor palumbarius*).

Of the falcons we can only refer to the game-falcons, the typical species of which, the peregrine falcon, is shown in our cut. This is a cosmopolitan species, found in almost every part of the world. None of the falcons exceed it in courage, daring, and rapacity; and only the gyr-falcon — a more northern species — was more highly esteemed in those days now long gone by, when falconry was a noble sport. It is a strong flier, and one which was started after a little bustard at Fontainebleau

FIG. 431. — Peregrine falcon, or duck-hawk (*Falco peregrinus*).

during the reign of Henry II. of France was taken the next morning at Malta, thirteen hundred miles away. The recognition of the individual was certain, for it wore on its leg the ring of the king. The peregrine feeds largely on sea-ducks, and from this derives its other name of duck-hawk. It lives to a great age. The story goes that one was caught at the Cape of Good Hope, in 1793, that wore a golden collar about its neck with the date 1610, and an inscription indicating that the bird formerly belonged to King James the First of England. Its eyes had lost their youthful lustre; and the feathers about its neck were white with age; and yet it appeared lively — very lively for a bird of a hundred and eighty odd years.

Before giving an account of falconry we must call the attention to the very curious bird, the caracara, or carrancha, depicted in the cut. It is an eagle or hawk, of uncertain position, but possibly should be placed nearest

FIG. 452. — Caracara eagle (*Polyborus tharos*).

the falcons. It ranges from our southwestern states south over Mexico, Central America, and Brazil. In its structure it is near the falcons; in habits it is a vulture, gathering in numbers like the turkey-buzzards about any carrion.

The peregrine and the gyrfalcon were the principal birds used in falconry; and these were restricted to the nobility and the king. The common people, if they wished to indulge in the sport, could only fly the kestrel or the sparrow-hawk.

One to-day can hardly imagine the prominence which falconry occupied for over a thousand years. " It was thought sufficient for noblemen's sons to wind the horn, and to carry the hawk fair, and leave study and learning to the children of meaner people "; and a German author of 1485 complains that "the gentry used to take the hawks and hounds to church with them, disturbing the devotions of those religiously inclined, by the screams and yells of the birds and beasts." All game and game-birds were scrupulously protected. Any one finding a hawk was compelled to carry it to the sheriff; and any attempt to conceal it was punished in the same way as stealing. Falcons commanded enormous prices; and at least one case is on record where a pair brought a thousand guineas. At the beginning of the sixteenth century hawking began to decline. Then any one might keep a hawk; and James the First wrote for the direction of his son: "As for hawking, I condemn it not; but I must praise it more sparingly because it neither resembles the warres so near as hunting doeth, in making a man hardie and skilfully ridden on all grounds, and is more uncertain and subject to all mischances; and which is worst of all, is there through an extreme stirrer-up of the passions." The spread of puritanical ideas and the establishment of the Commonwealth in Great Britain put an end to hawking; and it was in vain that Lord Orford, in the last century, spent a fortune in trying to revive an obsolete sport.

Not only in England and France did falcons hold a high place, but all the regions of the east; India, China, Tartary, and the like, all had their game falcons, and many continue the sport to this day.

This once popular sport demands a brief description; but there is no need to introduce here the details nor any of the technical terms used in books, which amount to hundreds.

The falcons, before being used in the chase, had to undergo a long and careful training, and the master of the hawks held a high place. The training begins with accustoming the bird to the human being and teaching him to recognize his master's voice. At first he is fed from the hand, and then, with a string attached to the leg, is allowed to fly for his food, until at last the bird is allowed to fly freely and to seek its own quarry, but still care is taken that it shall take only the sort of game which it is intended to chase.

A hawking-party presented a gay sight. Ladies and gentlemen went to the appointed place, the master of the hawks and his servants carrying the birds, each with a hood upon his head, while the master of the hounds followed with his pack. The hawk was set loose, and it quickly mounted upwards, and with a circling flight, kept his eye open for game. The dogs were now freed from the leash, and soon started up the birds from the

woods and underbrush. Some of the birds thus flushed afforded but little
sport ; but when a heron was driven out, the excitement rose high, for no
bird, excepting the kite, promised a finer spectacle. Both birds rose in
circles, those of the heron being small, those of the falcon large, each trying to rise above the other. Up and up they went, until they became mere
specks : but at last blood tells : the falcon gets above the quarry, and,
swooping down, strikes it with all his force. The mere shock of the collision may kill a smaller bird, but not so with a heron. Still, its strength
is gone, and it falls to earth, bringing with it the falcon, whose talons are
buried in its back.

Wise as an owl is a well-known phrase. Owls certainly do have a wise
look. The fluffy plumage of the head, and the large eyes, with their
strange setting of radiating feathers, all take part in giving this appearance of great erudition. What, for instance, in the whole range of nature,
has more the appearance of judicial gravity, critical acumen, and deep
thought, than the barn-owl whose image is before you ? No judge upon
the bench, no professor behind his desk, can begin to equal this bird of
wisdom. Still, it is a well-known truism that appearances are often
deceitful. The dignified judge often has nothing but his dignity : too
many of our professors profess to know, and stop there ; and the owls, on
their part, are content with seeming wise without the reality, either intui-
tive or acquired. But what difference does it make ? People and things
are taken for what they seem, and not what they really are. 'Feather-
top,' of Hawthorne's tale, captivated little Polly Gookin, and Minerva, the
Goddess of Wisdom, was so imposed upon by one of the owls that she
selected it as her confidential adviser. Who can tell but what some of
the foolish knowledge of ancient Greece and Rome had its origin in some
inanity which this pretentious bird whispered in the ear of the credulous
goddess ?

Most of the owls are nocturnal birds, keeping themselves concealed during the day, and only at evening beginning their flight. Owing to the dif-
ficulties of observation comparatively little is known of their habits. We
can easily see the eagle as it soars high in the air ; the water-birds take the
day for all their activities ; it is only during the hours of light that our
common birds display all their wonderful habits ; but the owls are pre-
eminently birds of the night. In the day-time their eyes seem to be of
little use, and then is the opportunity of all the other birds. When they
find an owl sleeping away the day, the scene is highly interesting. Then
all animosities but one are forgotten ; the loud shrieks constantly attract
more and more, all bent upon mobbing the owl. Round and round they
fly, each screeching its loudest, and each doing its best to torment the bird

of night until, driven to distraction, the owl seeks refuge in flight ; but in vain — the angry mob follows him sometimes for miles before they are satisfied.

At night the owl has its revenge. Wonderful plumage it has : so soft and fluffy that even in its most rapid flight it makes not the slightest noise. Then the large eyes see to perfection, and not a twig is stirred

FIG. 453. — Barn-owl (*Aluco flammeus*).

even in the densest underbrush while the bird is seeking its prey. If then it chance to run across one of its persecutors of twelve hours before, beak and talons make short work.

The barn-owl figured is found in the temperate parts of the world : Europe, North and South America, India, and Australia, all containing forms of this cosmopolitan species. With us it is abundant in the southern states and on the Pacific coast. Its voice is a harsh and rasping screech, and this bird far more deserves the name screech-owl than the bird which actually bears it. Its name barn-owl is not over-appropriate. It sometimes inhabits barns, but it does not confine itself to them; church towers, hollow trees, and even holes in sandy cliffs are occupied by it.

In the writer's estimation the most attractive of all our owls is the little Acadian, or saw-whet owl, a very close relative of the figure on the

Fig. 434. — Left, Tengmalm's saw-whet owl (*Nyctala tengmalmi*); right, pigmy-owl (*Glaucidium passerinum*).

left of Cut 434. Unfortunately it is not very common with us. Were it more abundant, it would be an advantage, as, like all the other small owls, it devours a large number of mice, moles, insects, and other vermin.

Our common screech-owl is a far more abundant bird, but its name is a misnomer. It does not screech nearly as much as do many of the other owls. It is strictly nocturnal, feeding, like the saw-whet, on mice and moles. Larger, but with the same curious tufts, are the great horned owls, which, however, take game of a larger size. These, together with the hawk-owls, the short-eared owls, and the barred owls, must be dismissed without further details; for the description of the habits of each would be but a presentation of a mass of facts uninteresting to any but the professional ornithologist. There are, however, one or two which possess sufficient individuality to make them noticeable here. First are the pigmy-owls, one of which is shown in the right of Figure 434. With one or two exceptions it is the smallest of the owls, measuring but about seven inches in length. It is one of the most diurnal members of the group of owls. Then there is the great snowy-owl, which every few winters comes down from the north in considerable numbers, seemingly for the sole purpose of offering themselves as sacrifices on the bench of the taxidermist.

Fig. 435. — Screech-owl (*Scops asio*).

The burrowing-owl of our western territories needs more mention. On page 389 will be found a quotation from the pages of Dr. Coues as to the happy family formed by the rattlesnake, the prairie-dog, and this absurd-looking, long-legged owl. There is no use of repeating the refutation of the time-worn story of how these animals live together in a most perfect accord. They do nothing of the kind: the owl selects the burrows of the prairie-dog merely because they furnish eligible homes, and frequently both owl and dog will occupy the same burrow. Still far more

Fig. 436. — Burrowing-owl (*Speotyto cunicularia*).

frequently the owls will be found in a part of the village deserted by the dogs, and it is not impossible that the former have evicted the rightful tenants.

Not only do these burrowing-owls occur in our western territories; they are found in Florida, the West Indies, and in South America; everywhere using subterranean burrows in which to raise their young. Where the prairie-dogs are lacking, the burrows of other animals — foxes, badgers, squirrels, and the like — are used, and in South America the holes of the Patagonian cavy are employed for this purpose. There is a difference to be observed in the habits of the burrowing-owls of North and South America. In the latter country they are, like most owls, nocturnal, but with us they go about by day. As one approaches one of the villages of the prairie-dogs, these owls present a most curious spectacle. They stand bowing and ogling and turning the head in the most grotesque manner. This doubtless arises from the nature of the eyes, the head being put in various positions in the effort to see the intruder more distinctly.

PARROTS.

There is scarcely a group of birds which possesses more interest than do the parrots, and yet the interest is of such a character that no description will do justice to it. It is strictly personal, and depends partly upon the bright colors, partly on the affectionate disposition, but far more upon the wonderful conversational powers of these birds. Man, it is often said, is the only talking animal, and hence it is that when we see anything possessing the power of articulation, — be it Faber's mechanical man, a doll which utters 'mamma' when the stomach is pinched, or a raven or parrot with its innate power of reproducing human speech, — our interest is at once aroused.

The strangest of the parrots inhabit that strange corner of the earth, Australia and New Zealand. In the latter country lives the owl-parrot, or kakapoo, a beautifully marked green, yellow, and black bird, which in many respects combines the features of the owls and parrots. It has a number of stiff hair-like feathers around the eye, which form facial discs, or goggles, which at once recall the owls; while its burrowing habits suggest the last owl that was mentioned above. This bird rarely flies, and then but weakly, and one kept in London, when especially pleased, used to "march about with its head twisted round, and its beak in the air, wishing, I suppose, to see how things look wrong way up, or perhaps it wished to fancy itself in New Zealand again." The ground-parrots of Australia are said to be similar in habits.

In every aviary and zoological garden may be found one or more cockatoos, welcoming the visitor with their cry of 'cock-a-too,' or in mock or real anger chasing their fellows about with screams, which form the great-

est contrast to the tones with which they pronounce the word "pretty
cocky." In a group where all are beautiful it is difficult to award the prize
of beauty. That experiment was tried long ago; and if you wish to know
the result, turn to Homer. Certainly without any chance of producing the
apples of discord, we may say that the rose-crested cockatoo — white all

Fig. 437. — Rose-crested cockatoo (*Plictolophus mollaccnsis*).

over, except the rose-colored crest — and its near relative, the yellow-
crested form, are among the most beautiful of birds. In a cage one may
watch their motions for hours without tiring. At one moment quiet as
can be, they are climbing about their perches and the bars of their prison,
as one might say by teeth and toe-nail, using their beak like another foot,
and then without the slightest apparent cause, they are all excitement and
anger. Every feather is erect, while the magnificent crest is opened and

closed in the most wonderful way; an instant more, and all is calm as before.

The great black cockatoo is not so handsome a bird; but it is exceedingly interesting in its habits, as described by Mr. Wallace. It feeds upon various seeds; but the kernel of the kanary-nut seems to be its favorite food. "The shell of this nut is so excessively hard, that only a heavy hammer will crack it; it is somewhat triangular, and the outside is quite smooth. The manner in which the bird opens these nuts is very curious. Taking one endways in its bill, and keeping it firm by a pressure of the tongue, it cuts a transverse notch by a lateral sawing motion of the sharp-edged lower mandible. This done, it takes hold of the nut with its foot, and, biting off a piece of leaf, retains it in the deep notch of the upper mandible, and again seizing the nut, which is prevented from slipping by the elastic tissue of the leaf, fixes the edge of the lower mandible in the notch, and by a powerful nip breaks off a piece of the shell. Again taking the nut in its claws, it inserts the long and very sharp point of the bill and picks out the kernel, which is seized hold of morsel by morsel by the extensile tongue. Thus every detail of form and structure in the extraordinary bill of this bird seems to have its use, and we may easily conceive that the black cockatoos have maintained themselves in competition with their more active and more numerous white allies, by their power of existing on a kind of food which no other bird is able to extract from its strong shell." The cockatoos all come from the eastern seas.

The lories, the rosilla parakeets, the pigmy-parrots (scarcely larger than one of our warblers), are all interesting and beautiful; while the little hanging or bat-parrots are curious from the way in which they sleep. Instead of perching like any other well-behaved bird, they grasp a limb with one foot just as does a bat, and hanging head downward, thus take their rest. The affectionate little love-birds are also common pets, and are admired by their owners not only for their bright colors, — green prevailing, — but for their disposition as well. Such love as each evinces for its mate is rarely equalled among the birds.

The talking powers of the parrots reach their highest development in the Jackos, or gray parrots, of western Africa. Detailing the remarkable speeches of parrots is like giving the bright sayings of children; in most cases the brightness is perceived by only those interested; but in one case the remark was so apt that it demands repetition here. It was at a parrot show in England where a prize was offered for the best talker. Parrot after parrot was brought into the room by the expectant owners, their cages uncovered, while the judges recorded their opinions. At last a gray parrot was brought in. He looked around the room, and then exclaimed,

"By Jove! what a lot of parrots." It is hardly necessary to say that he was awarded the prize.

The macaws, of which there is a large number, are American birds, but birds with almost nothing attractive about them. Their voices are extremely harsh, and they never talk well. Bright colored they are, but

FIG. 438. — Red-faced love-bird (*Agapornis roseicollis*).

such colors and such arrangements of color! The lack of harmony between the different shades could not be worse were they to run loose in a painter's shop, and to tumble into one pot of paint after another. Here is a patch of blue, and right beside it is a large daub of red, and then comes green and yellow. The whole is an outrage to every artistic sense, but doubtless the birds do not regard it as so horrible.

MACAWS (Arara).

Our Carolina parakeet is the only parrot in the United States, and this is apparently on the road to extinction. Years ago it had a wide range, and was abundant in localities where for fifty years not an individual has been seen. It was always a southern bird, but in 1790 a flock appeared in January near Albany, N.Y., creating the greatest consternation among the inhabitants, who believed it the portent of some disaster. They have appeared in Pennsylvania, Ohio, Indiana, Illinois, and Iowa, in some in-

Fig. 439. — Carolina parakeet (*Conurus carolinensis*).

stances within very recent years, but on our east coast their range has been greatly limited. To-day the name Carolina parakeet is a misnomer; for they hardly occur at all in that state where once they were so abundant. It is still abundant in Florida, but its fate there is a mere question of time. Mr. Allen in 1871 wrote: "Hundreds are captured every winter on the lower St. Johns by professional bird-catchers, and sent to northern cities. Thousands of others are destroyed wantonly by sportsmen. Concerning the needless slaughter Mr. Boardman thus writes: · The little parakeet must soon be exterminated. One Enterprise party would sometimes shoot forty or fifty at a few discharges for sport, as they hover about when any are shot until the whole flock is destroyed.' From its habit of feeding upon the tender maize in autumn, it is somewhat injurious to the farmer, and for this cause, also, many are killed. It is also more or less hunted as a game-bird. It is well known that the parakeet formerly inhabited large portions of the United States where it is now never seen, and the cause of its disappearance has been deemed a mystery. Such facts as these, however, seem to render clear what its ultimate fate must be in the United States — extermination."

In color this bird is green, shading off into a yellow below; its forehead is brick-red, and the rest of its head and its neck yellow. Besides cultivated grain, it is very fond of the seed of the cockle-burr. Though it occurs in our limits, we really know far less of its habits than we do of many rarer or far more distant forms. All of the parrots, however, are cunning, and replace the monkeys among the birds. They go in large flocks, and are eminently gregarious.

As among other groups of birds, so among the parrots, species have been exterminated within comparatively few years. One species, for instance, was found on but a single island of only about five square miles' area; this is now gone. So with the Madagascar parrot. Even in this century it was brought alive to Europe, but to-day it is extinct. Specimens are even more rare than of the great auk, the Labrador duck, or Pallas's cormorant; for they are only to be found in the museums of Paris and Vienna.

PICARIAN BIRDS.

For the large assemblage of birds now to be taken up, there is no collective common name; the one chosen is the nearest approach to one, and implies that all the forms included bear a more or less distinct resemblance to a woodpecker. The group contains nearly two thousand species; we can afford but little space to them, and hence think it best to largely confine our remarks to the most interesting and foreign forms, leaving the more commonplace and many of our native birds unmentioned, except as they are represented by cuts. These remarks will also apply to the next group. — the Passeres. This, however, is the less to be regretted since books abound treating of these forms, while the birds already enumerated have been more neglected in popular works.

First in order come the cuckoo-like forms, and of these attention must first be directed to the plantain-eaters and touracous, all of which are confined to Africa. The most remarkable feature about these birds is the green coloring-matter in their feathers. In all other birds the green color is produced by the structure of the feathers in much the same way as the color in a pearl, as indicated in an earlier part of this work (p. 79). Here, and here only, a green pigment occurs, and what makes it more

FIG. 440. — Ivory-billed woodpecker.

strange is that if these birds wish to preserve their beauty, they must stay in doors when it rains, as the water washes out the color.

The cuckoos of America are often attributed with traits which do not belong to them, and this arises, not from any intention, but from assigning to them the character of the far more celebrated cuckoo of the Old World. And even that has not escaped. Many are the superstitions that hang about it. In Europe you will be told that it turns into a hawk in the fall, that it can lay eggs of any color it wishes, and numberless other tales, but none quite so marvellous as these; but all about equally well founded. Possibly the most interesting of all the actual facts in the life of the European bird is its habit of laying its eggs in the nests of other birds, and thenceforth leaving them to their fate. In the choice of the foster-mother she is not very particular, but takes a nest where she can readily obtain

Fig. 441. — American yellow-billed cuckoo (*Coccygus americanus*).

access. Equally remarkable is the voracity, nay, even gluttony, of these birds. Mr. Cuckoo, when he begins to eat, does not know when to stop. All those little endearments which other bird-husbands practise are neglected by him; and even when that important time of courtship comes round, it is Miss Cuckoo that has to do the courting and to pop the question, for her lord that is to be is too busy in filling his stomach to devote much time to such affairs.

Our American cuckoos, of which we have three, are not so lazy as are their cousins of Europe; they hatch their own eggs, but no one ever heard them complain about the competition with the pauper labor of Europe. Perhaps they don't dare to; for they are about the most cowardly of all our birds, a robin or a sparrow putting them to flight.

A fourth species in the United States is the road-runner, chapparal-cock, or ground-cuckoo, which reaches from Texas to southern California, and south into Mexico. It is a queer-looking bird, quick and active in its

Fig. 442. — Chapparal-cock, or road-runner (*Geococcyx californianus*).

movements, and delighting in the mountainous and desert regions, or in those where the vegetation is scrubby and consists largely of cactus. A road or path has peculiar attractions for it, and it will run rapidly along on it, hunting for insects, lizards, and other articles of food. It is said that it sometimes kills and eats rattlesnakes, but this is doubtful.

The ani, or savanna blackbird, is just entitled to be called an inhabitant of the United States, as it enters Florida. It is, however, seen at its

best in the West Indies.   The name Crotophaga means tick-eater, and is
well applied to these birds; for, like some other birds, it eats the parasites
from the backs of the cattle, leaping on the backs of the cows while they
are grazing, and eagerly picking off all the insects: when the back is
cleaned, they take the tail, and clinging to it with their strong claws, pick
everything edible out of the terminal tuft of hair.   Again, they will seek
their food in the earth, and here the large bill, knife-edged above, comes
into play, for it cuts like a ploughshare.   In flight the bird presents a

FIG. 413. — Savanna blackbird (*Crotophaga ani*).

peculiar appearance.   The body is held perfectly straight, the tail being in
the same line, so that, viewed from the side, it has the aspect " rather of a
fish than of a bird."

Most remarkable is its nesting habits.   Sometimes one bird will make
its own nest; but usually a company unite, and build an immense nest of
basket-work, usually on a high tree.   This is used in common by all the
flock, half a dozen or more laying their eggs together, and all assisting in
the incubation, and rearing the young together.

The group of goat-suckers and their allies play as important a part in
the superstitious beliefs of the uneducated, as they do in the realms of

nature. Yet why they should be thus maligned is a question no one has yet answered. If a whippoorwill but sound the notes which give it its name upon the door-sill of a house, one of its inmates is soon to die a sudden death. The voice of the night-hawk has the same evil portent; while the white spots upon its wings are firmly asserted to be silver dollars. The goat-sucker, too, derives its name from a belief that it sucks the milk of goats and cows. "Poor little injured bird of night," says Waterton, "how sadly hast thou suffered, and how foul a stain has inattention to facts put upon thy character! Thou hast never robbed man of any part of his property, nor deprived the kid of a drop of milk." This goat-sucking belief has, however, an explanation. As the shades of evening fall, the

Fig. 444. — Night-hawk (*Chordeiles virginianus*).

goat-sucker flies around the herd, jumping every now and then up to their bellies. It is not milk that it is after, but the insects that are tormenting the flock.

This subject of superstitions connected with birds is a large and interesting one. So queer are some of the beliefs that it seems almost impossible to trace them to their source, while others are as easily explained as is the goat-sucking propensity of the bird just named. Here is a splendid chance for some Max Müller or Grimm to work out the origin of a wonderful folk-lore. But the study will have one aggravation; for the investigator will find that people, apparently well informed, stick with the greatest pertinacity to their beliefs, and will stoutly affirm that they have seen the swallows dive beneath the surface of the pond in the fall, and emerge again in the spring.   If you intimate that possibly they have made a mistake, they give you a reply, which while you believe — yes, know — to be entirely false, you cannot reply to without casting the most direct and unmistakable imputations upon their honesty.   They will tell you that they have killed goat-suckers and found the milk in their stomachs; and that when fishing through the ice, the swallows have been brought up on their hooks alive, but in a state of stupor.

To me there is something very interesting and attractive in the appearance of any of these nocturnal birds.   The soft and fluffy character of the plumage, its beautifully mottled sober colors, the bright eyes, the short, broad beak, and the hairs which fringe the gape, all add to the appearance.   And then there is the note.   The night-hawk at the pairing time goes through the most wonderful evolutions.   At dusk he rises high in the air, and then, hawk-like, he closes or half closes his wings and darts down to the earth, making a most curious booming sound.   Again he mounts and goes through the same evolutions.   This note is difficult to describe; possibly it is best appreciated by likening it to blowing into the bung-hole of an empty barrel.   This is not a true voice; but is produced by the rush of the air between the quills of the wings.

Then our whippoorwill's note is always pleasing, as in the evening it comes from the trees.   It tells you that Will has done something wrong, that he must be punished; but in such a tone that one knows the bird is sorry for the misdeeds of William.   Others express only sympathy, and do not ask for the punishment; they only say 'poor-will,' but in such a plaintive tone that one can hardly help thinking that their hearts are breaking with grief.

The bee-eaters of the Old World have a curious diet, eating bees, wasps, and other insects, which they seize as they fly past.   They are brightly colored forms, which dig deep holes in sandy cliffs, in which to build their nests.   Strange it is that their diet does not affect them seriously, for they must frequently get stung by their victims; and the sting of a bee or wasp kills many other birds of the same size, and it is even stated that a duck will not survive an injury of this sort.

In tropical America the bee-eaters are represented by the sawbills, or motmots, the greatest peculiarity of which is found in the middle tail-feathers.   These lack the web on either side for a considerable portion of their length, but on the end it remains, giving the whole the appearance of a paddle.   Various are the theories of naturalists and natives as to the cause of this.   Some say that the bird, when on the nest, keeps its tail in constant motion, and thus by rubbing on the edge of the nest the mutilation is produced; but this certainly cannot be the case; for if it were, the shaft

would show signs of abrasion, which it never does. It would appear probable from what various observers state, that in many instances, at least,
the bird does the pruning with her saw-like bill; but why, is another
question. All of the motmots live beneath the surface of the earth, some
using the burrows of armadillos, others in caves, while wells are utilized
by still others. Mr. Gaumler says : —

" About twenty of these birds lived in a well from which I used to draw
water every day. The well was about forty feet deep, had been cut through

FIG. 445. — Whippoorwill (*Antrostomus vociferus*).

a porous shell-limestone, and its walls contained many cavities into which
a man could crawl many feet, but was obliged to back out. I have frequently gone many yards into these caverns to investigate the home of
the sawbills and their work therein, and have always come out feeling well
repaid for all the danger, having invariably seen something new and
interesting. I have found the young birds in almost every stage of
development — those with the tail-feathers just starting being the most
interesting. The feathers all seem to grow alike to a certain point,
except the middle ones, which are always a little broader towards the

end; there all cease to grow except the two middle ones, which soon pass
the others by about an inch and a half.   Up to this point the webs of
these two feathers are just the same throughout, except the sub-terminal
portion, which is much narrower.   Thus far no mutilation has taken place;
but as soon as these feathers exceed the others a little more, the web

Fig. 446. — Kingfisher (*Alcedo ispida*).

begins to disappear, and the outer web of each feather is generally taken
off first."

     The kingfishers, of which we know about a hundred and fifty species,
are brightly colored birds, the typical members of which are as fond of
fish, and as dextrous in catching them, as any fish-hawk, or any disciple
of Isaac Walton.   These fishing forms are too well known to need any
description.   Every one has seen them perched on some branch overhang-

ing the river, watching the shoals of little fishes below, and every now and then making a headlong plunge into the water, and almost instantly emerging with a fish in its beak. All of the typical kingfishers do not feed on fish. Wallace instances one in the island of Lombock, which belies its name. It does not frequent the water, does not live on fish, but seeks its dinner of insects, centipedes, and molluses in the dense, damp thickets of that tropical isle.

Of the less typical kingfishers we merely mention the racket-tailed king-fishers of the Malay Archipelago, with the tail-feathers much like those of the motmots; and then proceed to give a quotation from the pages of Mr. Wheelwright, concerning the 'laughing jackass' of Australia : —

"About an hour before sunrise the bushman is awakened by the most discordant sounds, as if a troop of fiends were shouting, whooping, and laughing around him in one wild chorus. This is the morning song of the laughing jackass, warning its feathered mates that daybreak is at hand. At noon the same wild laugh is heard; and, as the sun sinks into the west, it rings again through the forest. I shall never forget the first night I slept in the open bush in this country. It was in the Black Forest. I awoke about daybreak, after a confused sleep, and for some minutes I could not remember where I was, such were the extraordinary sounds that greeted my ears; the fiendish laugh of the jackass, the clear, flute-like notes of the magpie, the hoarse cacks of the wattle-birds, the jargon of flocks of leather-heads, and the screaming of thousands of parrots as they dashed through the forests, all joining chorus, formed one of the most extraordinary concerts I have ever heard, and seemed at the moment to have been gotten up for the purpose of welcoming the stranger to this land of wonders on that eventful morning. I have heard it hundreds of times since, but never with the same feelings that I listened to it then. The laughing jackass is the bushman's clock, and, being by no means shy, of a companionable nature, a constant attendant about the bush-tent, and a destroyer of snakes, is regarded, like the robin at home, as a sacred bird in the Australian forests. It is an uncouth-looking bird, a huge species of land-kingfisher, nearly the size of a crow, of a rich chestnut-brown and dirty white color; the wings slightly checkered with light blue, after the manner of the British jay; tail-feathers long, rather pointed, and barred with brown. It has the foot of the kingfisher, a very formidable, long, pointed beak, and a large mouth. It has also a kind of crest which it erects when angry or frightened; and this gives it a very ferocious appearance."

Strange birds are the hornbills, one of which is figured here. They get their name, as one may readily see, from the immense bill which bears

on its upper surface a curious casque, or helmet, varying in shape and size
with the species. Though so large, this bill is very light, and is of a
regular honey-combed structure; but this is much more marked in some
than in others. In the great hornbill it is very much heavier than in
others, and this bird beats the trees with its heavy hammer-fronted casque,

Fig. 447. — Hornbill (*Buceros bicornis*).

producing resounding thuds which can be heard long distances. Before
this hammering habit was actually known, Professor Flower was confident
that the head was used as a pounding organ, so peculiar is its structure.
The bone in the part that hammers is very thick, and from it proceeds a
layer of dense bone which passes above the cavity of the brain, so that the
strain comes on the neck, while the brain itself is out of the line of the
shock.

The breeding habits of these birds are as strange as is their appearance. A short quotation from A. R. Wallace is given here, and some farther remarks will show the essential points : —

"I had sent my hunters to shoot, and while I was at breakfast they returned, bringing me a fine *Buceros bicornis*, which one of them assured me he had shot while feeding the female, which was shut up in a hole in a tree. I had often read of this curious habit, and immediately returned to the place, accompanied by several of the natives. After crossing a stream and a bog, we found a large tree leaning over some water, and on its lower side, at a height of about twenty feet, appeared a small hole, and what looked like a quantity of mud, which I was assured had been used in stopping up the large hole. After a while we heard the harsh cry of a bird inside, and could see the white extremity of its beak put out. I offered a rupee to any one who would go up and get out the bird, with the egg or young one, but they all declared it was too difficult, and they were afraid to try. I therefore very reluctantly came away. In about an hour afterward, much to my surprise, a tremendous, loud, hoarse screaming was heard, and the bird was brought to me, together with a young one which had been found in the hole. This was a most curious object, as large as a pigeon, but without a particle of plumage on any part of it. It was exceedingly soft, and with a semi-transparent skin, so that it looked more like a bag of jelly, with head and feet stuck on, than like a real bird."

To this a few explanatory remarks may be appended. The female bird, as the breeding time draws near, takes her position in some hollow tree, and then both birds proceed to wall up the opening with mud and ordure, leaving a narrow slit-like opening through which the female receives all her food of fruit which the male regularly brings to her. If incubation has progressed far, the female is a sorry-looking sight when taken from her hole. She is wasted and dirty, and her wings are so stiff that she cannot fly. It is difficult to see how such a habit could have arisen. The advantages gained seem very slight, and one can hardly realize that they counterbalance the danger to the female if the male should be killed. The hornbills all belong to the Old World ; in size they vary between a raven and a jay.

In the forests of tropical America the toucans seem to take the place of the hornbills of the Old World ; at least so far as large bill and fruit-eating habits go. They, too, make their nests in hollow trees, but they do not practise that plastering art of the hornbills, which makes one recall those weird tales of some culprit, or, it may be, rival lover, of the Middle Ages, being walled up alive in some vault. Exactly why these birds have such.

large bills it is not easy to see. The story goes that, after plucking the fruit on which they live, they toss it into the air, and their large bills are of use in catching it as it falls, and directing it in the proper course down the throat: but trustworthy travelers deny the existence of this habit, and say that they merely throw back the head and let the morsel roll down into the gullet. In their native forests the toucans go in large flocks, and

FIG. 448. — Toucan (*Ramphastos picatus*).

their bright colors, like those of many other tropical birds, go far towards enlivening a landscape conspicuous from the scarcity of bright flowers. When they feed, a sentinel is posted; and when danger is near, he gives the shrill, yelping, warning note *tucano*, from which the birds derive their name.

The honey-guides of Africa and the East Indies derive their name from a peculiarity in their habits. They are small birds, which are said to guide men to stores of wild honey. Strange, and almost fabulous, as this seems, the story is well authenticated. One writer gives an account which we here condense. The bird had been flying almost in the faces of his party as they were making a journey in the Transvaal. The negro boys were paying it considerable attention, and at last the head of the party learned that this was the far-famed honey-bird. As soon as the oxen were outspanned, the whole party set out in pursuit of the bird. It led them on

FIG. 449. — Three-toed woodpeckers (*Picoides tridactylus*) of Europe.

from tree to tree for nearly a mile, and at last flew exactly to a tree, where, after a little looking, the party found traces of a nest of bees. One of the party climbed the tree, and lighting a bunch of dried grass, struck the trunk. Out flew the bees, and their wings were quickly singed, and then a few blows of the axe laid bare the comb, a piece of which was left on a bush as a reward for the bird.

The woodpeckers, the central members of the picarian birds, are a host in themselves, some three hundred and fifty strong. Their habits are well known and are essentially similar throughout the whole series. All pick the wood in their search for larvæ; all have the stiffened tail-feathers to aid in holding them in position, while their chisel-like beak is boring the holes in the bark. The question whether their habits are to be regarded for good or for evil to the farmer has often been raised, but it is hard to say in which direction the balance lies. They do good by destroying injurious insects; but, on the other hand, there can be no doubt but that their punctures affect the tree, especially at the time when the sap is flowing. Some are even accused of eating the soft inner bark.

There is one structural feature in the woodpeckers to which attention should be directed. In all birds (and in mammals as well) the tongue is attached to a little Y-shaped bone, called the hyoid bone, because some old anatomist with a very vivid imagination imagined that the same bone in man resembled the Greek letter Y. Now the woodpeckers have to get their food from the holes which they make in the bark; and when one of these holes strikes the burrow of some insect, in goes the tongue, probing it to its depth, and bringing out the grub. To do this it is necessary that the tongue be very extensile, and this is rendered possible by having the hyoid bone very long, and running,

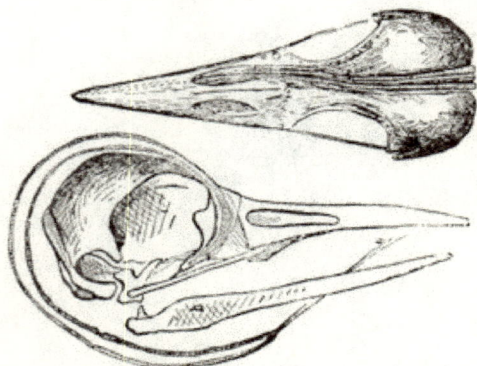

Fig. 450.—Skulls of woodpeckers, from above and from the side, showing the tongue and the hyoid bones curving around the base of the skull and passing above into the nostril (above) and around the orbit (below).

as shown in the cut, around the base of the skull and over its crown. The tongue itself is furnished with barbs like those of a fish-hook, thus rendering it possible to draw out the grub speared by it.

The trogons, which live in the tropical parts of both continents, are beautiful birds, the form figured (the quesal) being the handsomest of all. The male is by far the most beautiful of the two: it is of a rich bronze-green, which no humming-bird can excel, while the true tail — not the long tail-coverts — is black and white, and the belly is a rich vermilion or crimson. Is it any wonder that Guatemalans have adopted it as their national emblem, and placed its portrait on their postage stamps? Its body is no larger than that of a crow, but the two longest feathers may hang down for nearly three feet. One would think that such long ornaments would incommode him in his flight through the forest: not so; he flies rapid and straight, his long banners streaming behind. It feeds almost solely upon fruit, and this it takes in a peculiar way. It does not perch on a limb and eat its fill, moving about to get more as soon as those within reach are exhausted; but flies from its perch, takes the fruit on the wing, and returns again to his resting-place, the whole flight being accomplished with a degree of elegance that cannot be described.

The best known to us of all the swifts is the chimney-swallow, which every student of ornithology will tell you is not a swallow at all. Still the name sticks, and will stick, in spite of all protests, and the best thing that naturalists can do is to gracefully submit to the inevitable. Our chimney-swallow is a native of this country, and before the days of houses it built its nest in hollow trees. Now it has entirely changed its habits in the settled part of our country, and taken entirely to chimneys. Inside of these the nests are built of twigs, glued together by the mucus secreted by the bird.

In another swift this glutinous secretion reaches a pecuniary importance; for it furnishes the celebrated edible birds'-nests of the Chinese markets, about which until recently but little has been really known. Most of the nests are gathered on the northern coast of Borneo, where the swifts make their homes in caves. The nest-gatherers, by the light of a torch, pull the nests off the rocky walls with a four-tined fork fixed on a large pole. Then the nests are sorted,—the best tied together with rattan, the poorer strung together. It takes about thirty of the best nests to make a pound, and these are worth about seven dollars a pound. The nests, too, have been subjected to a chemical and micro-scopical examination, and it is found that it was not of a vegetable nature, but consisted

Fig. 451. — Quesal, or paradise trogon (*Pheromacrus mocinno*), male and female.

almost solely of strings of mucine plastered together; and the probability is that this, like the 'glue' of the chimney-swallow, is secreted by the glands in the mouth of the bird.

The whole interest that surrounds those gems of the air, the humming-birds, has its centre in their beauty. One who has lived long in their home thus writes of them : " Humming-birds are unlike other birds in their mental qualities, resembling insects in this respect rather than warm-blooded vertebrate animals. The want of expression in their eyes, the small degree of versatility in their actions, the quickness and precision of their movements, are all so many points of resemblance between them and

FIG. 452. — Coquette humming-bird (*Lophornis ornata*).

insects." And yet they are beautiful enough to make up for many deficiencies of intellect, and one cannot stand before a case of them gleaming with ruby, turquoise, gold, — all the hues of the rainbow, and many others which Iris never dreamed of, — and think of anything except the magnificence.

The humming-birds all belong to the New World: a few enter the United States; but to see them in all their beauty one must visit the tropics. In all, about four hundred species are known. Excepting their plumage there is little about them to attract; they have no winning ways, their habits and actions seem almost automatic, they exhibit no intelligence, and even their fearlessness seems to arise from the want of a brain sufficient to appreciate what fear is. Then, too, they are very quarrelsome among themselves, and every traveler in South America tells of their duels, —duels between mites it is true, but combats which far more frequently terminate disastrously than they do with German students, or even in our own southern states. Yet stupid and uninteresting as they seem, every one will make an exception in their favor when thinking of the old adage, 'fine feathers do not make fine birds.'

And then such dainty little nests as they make. One may hunt round and round a tree or a bush where it is certain they have their home, and yet this little bit of architecture will almost invariably escape him. It is covered on the outside with lichens, so that it looks like nothing but a knot or small excrescence; but within it is a mass of down stolen from all the thistles in the neighborhood, and the whole so deftly put together with cobwebs, as to make it a fit setting for the jewel that occupies it, and the eggs, scarcely larger than pearls, that run no danger of breakage in its downy depths.

Many attempts have been made to keep these birds in confinement, feeding them on honey and water, syrup, and the like: some succeed for a short time, but sooner or later there comes an accident, and these little bodies do not have life enough to survive much injury. Then, again, the food is not sufficient. It is a poor substitute for the nectar of flowers; and besides, these birds need insects, and these cannot well be supplied in sufficient quantities. As much bright sunlight as is possible is another ne-

Fig. 453. — Ruby-throat humming-bird (*Trochilus colubris*).

cessity; and no human hand, no matter how skilful, careful, and thoughtful, can supply these requisites which nature offers freely.

PASSERES.

As 'Picarian birds' was defined as woodpecker-like birds, so the above heading may be translated by the words sparrow-like birds. Of these the numbers are legion. The latest catalogue of the birds of the world enumerated over twelve thousand species, and of these a half belonged to the present group. Our space will admit but an insignificant fraction of these, and hence only those most interesting will be mentioned at all. A large volume would be necessary: a few pages is all that we can afford. There is, however, one point to be noted: in many of the groups there are many species alike in structure and in habits, the only distinctions between them being of that character which only interests the ornithologist, and hence a description of one of the group is almost equivalent in a popular work to a description of all.

The first of these birds is the lyre-bird of Australia, a form which possesses no especial beauty, or interest, except that furnished by its unrivalled tail. Taking this into consideration, the bird well deserves its specific name, 'superba.' Of its habits comparatively little is known. It lives in the dense brush, and is very shy and wary in its habits. It is said to be an excellent mimic, repeating the notes of many other animals in the same way that does our mocking-bird. The body is scarcely larger than that of a bantam fowl, and the male alone, as in so many other cases, possesses the wonderful lyre-shaped ornament.

Fig. 454. — Australian lyre-bird (*Menura superba*), male and female.

Of the birds which in their structure resemble our tyrant-birds the ground-thrushes (which are not thrushes at all) must be mentioned on account of their great beauty. With one African exception all belong to the East Indies. No description can do the slightest justice to their beauty : that requires the painter's palate. Blues, crimsons, greens, yellows, and purples are lavishly used on their soft and fluffy feathers. They are wary birds, living in the woods, picking up their diet of worms from the ground, and turning to the dense underbrush on the slightest alarm. They do not appear to hurry, but they actually make good progress with their hopping gait, and it takes a skilful hunter to bag them.

The ground-thrushes are not nearly so pugnacious as their relatives the tyrant-birds; these latter are well exemplified by our familiar king-bird. This is a most irascible little fellow ; it is a regular tyrant, especially at the breeding season. Hawks and crows are his especial dislike. Let one appear in the neighborhood of his nest, and one soon sees how he is 'king' among birds. At first there is a sharp, twittering note, and then he launches himself at his enemy. Over it, under it, and all around it he flies, striking at it with all his might. The crow takes to its wings in the most ignominious manner, and flees from its persecutor. In this way this bird does good ; but still he has his sins. A special

Fig. 455.—King-bird (*Tyrannus tyrannus*).

luxury is furnished for his table in the shape of bees. He does not fear their stings, but makes sad havoc with these busy honey-gatherers.

Among the relatives of the king-bird must be mentioned the numerous fly-catchers and the phœbe-bird of our own climes — forms too common to need farther attention.

The cock-of-the-rock of South America, an orange-colored bird of South America, which has a curious, narrow crest upon the head, is a more distant relative of the tyrant-birds. It is a celebrated bird, and usually spoken of as showy and beautiful. The writer, however, never could see any especial claims to beauty in these birds. They look awkward and ungainly. Possibly the fault was with the taxidermist.

A rare and curious inhabitant of the upper Amazon forests is the umbrella-bird, which in size, color, and appearance is much like a crow, except for a crest of long, curved feathers, which, when elevated, turn forward over the head like a parasol. From the neck there depends a long bunch of steel-blue feathers, supported on a fleshy lobe. This lobe is connected with the vocal organs, and doubtless plays a part in producing the loud, deep, and flute-like note of the species. Says Mr. Bates: " We had the good luck, after remaining quiet for a short time, to hear its perform- ance. It drew itself up on its perch, spread widely the umbrella-formed crest, dilated and waved its glossy breast-lappet, and then, in giving vent to its loud, piping note, bowed its head slowly forward." This species figures in science as *Cephalopterus ornatus* — a most appropriate name.

Allies of this are the bell-bird, figured in our cut, and its two cousins. The naked-throated bell-bird is the least striking of the three; one of the

FIG. 456. — Naked-throated bell-bird (*Casmarhynchos nudicollis*).

others bears a large conical appendage upon the base of its bill, which it is capable of erecting, so that, like a miniature church-spire, it reaches a length of three inches, the whole surface being covered with white, star- like feathers, between which the black skin shows through. This gives the bird about as strange an appearance as one can imagine. The name bell-bird is derived from the note of the bird. It is very clear and loud, and can be heard for a long distance in the dense forests of Brazil, and it does not require a very vivid imagination to make the listener think that some little chapel must be near.

Farther south in South America live the oven-birds. These receive their name from their peculiar nests. These the bird builds of clay scraped from the neighboring puddles, firmly packed together, and dried in the sun, so that it is almost as hard as a brick. The nests are about a foot in diameter, and are shaped something like a spiral shell, and contains no soft lining, the eggs lying directly on the clayey bottom.

The skylark of Europe is one of the most celebrated of birds. Ornithologists and poets unite in their praise of its wondrous song. It rises from the earth in an almost vertical direction, singing as it flies, and it continues its song until it soars clear out of sight; and yet, if the day be still, one can hear the same song coming down from above, so powerful is it. Says Shelley : —

> "The pale purple even
>   Melts around thy flight;
> Like a star of heaven
>   In the broad daylight
> Thou art unseen, but yet I hear thy shrill delight."

Some years ago an attempt was made near New York to introduce the skylark into our country. Fortunately it was not successful; fortunately — for this bird with all its beauteous song would have proved but little less of a pest than does the English sparrow.

Our only representative of the larks is the shore-lark, with its curious 'horns,' or 'ears,' on the side of the head. It is a resident of our western states, and breeds there; but in the east they are seen only in the colder months, as they go north to Labrador and Newfoundland to rear their young.

We have just said that the shore-lark is the only representative of the larks in America; but there are certain of the pipits and wagtails which share the name, one even rejoicing in the name skylark. These are

Fig. 457. — Shore-lark (*Otocoris alpestris*).

mostly found around water, and have the habit of jerking their tails, when on the ground, in a most peculiar manner.

In almost every work on the east the bulbuls occupy a prominent place; the poets of India and Persia sing their praise just as their brothers of the west do those of the lark and the nightingale. And yet it would appear that their music is far inferior to that of the European birds mentioned. The bulbuls are sociable birds, and fond of staying in the neighborhood of man. Their song is a quick, cheerful chirruping warble, constantly repeated.

The thrushes and their relatives are eight hundred strong, and among them come some of the best of song-birds. Best known of all is the familiar robin, which comes each spring to build its nest of mud about our dwellings, to eat our berries and our cherries, and to do far more good than evil in destroying insects and worms by the thousand. Our robin is a fine singer, but he is far excelled in this line by his cousin, the wood-thrush. A quiet, retiring bird is this last. It has none of the familiarity of the redbreast; it cares nothing for the haunts of men, except to keep away from them; it loves rather the deep woods, and there, and there only, is its song heard. Early morn and early twilight are the times for the concerts; and such concerts — who can describe them? It is largely a concert of the flute; there is a tinkling little warble, long liquid notes, trills, and double-tonguing; now one bird taking it, and then his rival just beyond those bushes tries his throat. The whole is beautiful, though of a melancholy character.

Even more celebrated as songsters are the European nightingales. Would you know how they sing, or better, how they ought to sing? Then turn to the poets. Learn there that these birds sing only at night; that this peculiar melancholy sweetness of the song is caused by the bird leaning against a thorn; and then turn to nature and learn that all this is false. The bird needs not the inspiration derived from pain to give voice to its feelings; it can and does sing by day. It may be that it is overpraised, and yet the universal testimony of all — poets and naturalists — is that its song is unrivalled by any of the feathered choristers.

Relatives of the thrushes are the warblers of Europe, — how many there are only the ornithologist knows. They are far different birds from our warblers, and are much nearer our kinglets, one of which is figured here. These, however, must be dismissed, for there is another form, far more interesting, demanding attention.

Fig. 458. — Golden-crested kinglet (*Regulus satrapa*).

This is the celebrated tailor-bird of India and Burma, which sews together the leaves of trees to form a frame-work, or casing, for its nest. These birds select three or four leaves growing near each other, and proceed to sew their edges together so that a purse-shaped bag is formed. Inside of this the true nest — made of wool, vegetable fibres, and the like — is placed. For a needle the beak is used; this punctures the leaves, draws through the thread, and ties the knots. These birds do not, as is usually stated, sew dead leaves which they may pick up, but the dead leaves sometimes to be seen on the nests were alive when the case was formed, and have since died from the injuries caused by puncturing the holes.

The dipper, or water-ouzel, is a queer bird; for among a host of terrestrial relatives, it is an aquatic form. It only occurs in our western territories, in the neighborhood of clear streams. It hunts for its food beneath the water, swimming with its wings beneath the surface. Most interesting

Fig. 459. — Dipper, or water-ouzel (*Cinclus mexicanus*).

of all is its nest of moss, which is always placed near the water; sometimes so near that the spray keeps it moist. At other times, according to Mr. Stevenson, the bird supplies the moisture in the following manner: "One of the first things that attracted my attention was its manner of diving down into the water and then darting back and perching itself on the summit of its mound-like dwelling, where it would shake the water from its feathers and distribute it over the nest, apparently for the purpose of keeping the moss moist and in a growing condition." This sprinkling operation was repeated daily, and there seems every reason to believe that the sprinkling was for the purpose of keeping the outer coating of moss alive, and thus save the trouble of any repairs.

The wrens are familiar little busybodies, which need no introduction to any one. Some are exceedingly familiar with man, making their nests in his house or in his fence-posts, while others are more retiring; but all are cheery little bodies, and the little form figured comes at winter to sing his animated and pretty song.

FIG. 460. — Winter-wren (*Anorthoura troglodytes*).

In singing powers the mocking-bird is superior, and is very familiar to all from its living so well in confinement. His own song has a metallic sweetness, but to this it adds the notes of almost every other bird, and so perfectly that one cannot help being deceived by him. He is, in fact, a performer on the 'variety' stage; he gives you a medley of everything, sometimes introducing his notes in the most incongruous manner. A near relative of the mocking-bird is the cat-bird, familiar to all with its plaintive, sometimes almost exasperatingly reiterated 'mew,' exactly like that of the cat. But when it sings its own song, perched on some high tree, one can but wish that he would stick to that part of its performance and drop the rest. Its

FIG. 461. — Bank-swallow (*Clivicola riparia*).

voice is not so strong as that of the mocking-bird, but it is very melodious; and the bird is considerable of a mimic withal.

All of the swallows are much alike in both habits and structure. There is little chance of confusing them with any other group of birds, except with the swifts, already mentioned. Like them they have a strong, swift flight, and like them they live largely on the insects, which they capture on the wing. Most familiar are our barn-swallow, whose mud nests, plastered beneath the eaves of barns, are everywhere to be found; and the bank-swallow, or sand-martin, whose burrows are seen in high, sandy cliffs and river-banks. The elliptical hole leads to a nest some two or three feet from the surface.

The cedar-birds and the waxwings are really beautiful birds. They have crests on their heads, and, most striking of all, some of their wing-feathers are tipped with horn like red sealing-wax. The two look much alike, but the waxwing is the larger bird, as well as the rarest with us. It but rarely appears in the eastern United States, but in the west it sometimes occurs in enormous flocks. For years nothing was known of its nesting habits, except that it must bring forth its young at the far north. At last the eggs were found in Lapland by Mr. Wolley, who readily sold his duplicates for twenty-five dollars apiece. In America the nests have been found but twice, each containing but a single egg; one coming from

Fig. 462. — Bohemian waxwing (*Ampelis garrula*).

the Yukon, the other from the Anderson River. The waxwing is very erratic in its movements, and neither necessities of food nor weather seem to explain their wanderings. One year large flocks will visit a locality, and then for years none will be seen. This periodicity is far more marked in Europe than with us, and their advent in olden times was regarded as certain a foreboding of disaster as that of a comet.

The cedar-bird is much more familiar, and in the summer, when the cherries are ripe, this bird may be frequently seen upon the topmost branch, eating his fill. This cherry-eating habit makes the bird many enemies, but it far more than repays the damage it does, by the amount of insects and larvæ it eats. Only while the cherries and a few other fruits are ripe does it do damage. At other times it feeds upon insects of all sorts, and is especially fond of canker-worms.

Different in character are the shrikes, or butcher-birds. The latter name is especially appropriate for some of the larger forms, like the great northern shrike, which each autumn comes down from the north to set up his meat-market in our latitudes. It is a bold bird, with hooked bill, which, when no other bird is around, feeds upon insects; but if a mouse or small bird happens near, this furnishes the meal. They will frequently dart into a cluster of our English sparrows, kill two or three, and then proceed to hang the victims upon the thorns of various bushes, just as their human namesakes hang the quarters of beef

Fig. 465.—Great northern shrike, or butcher-bird (*Collurio borealis*).

or mutton in their markets. Some say that the butcher-bird does this so that it may have tainted meat; others, that it does it so as to have a stock to tide over some time of scarcity. The latter is the more probable view, especially if we add to it that the feet of the bird are very weak, and that thus it can hold the prey while pulling it to pieces with its beak. Sometimes the bird kills far more than is necessary to answer its wants; seemingly from mere cruelty and wantonness. In the Old World the common belief is that it kills nine animals a day, and hence it is called the 'nine-killer.' The long series of vireos, or greenlets, are to be merely mentioned as insect-eating allies of the butcher-birds, and some of them sing very sweetly. They are very common.

No more need be said of the industrious nut-hatches, which spend their time, like woodpeckers, digging the insects from the lichens and the crevices in the bark of trees.

The wonderful bower-birds of Australia and New Guinea, on the other hand, deserve more attention. The best-known species is that figured — the satin bower-bird. In itself it is far from striking; its colors are not gaudy, its shape is far from elegant. The bowers which they construct are the centre of attraction. These 'bowers' are apparently intended for courting-places, much like those of the argus pheasant already described, in purpose, but far more elaborate. This species makes an arbor-like gallery of twigs through which the male and female course with the greatest glee. The ground on which it stands, and the arbor itself, are tastefully decorated with all sorts of things, — shells, bones, leaves, feathers, and flowers. — some of which are brought from considerable distances. Once arranged, it is not allowed to stand; almost every day a new arrangement of the ornaments is introduced, the whole reminding one of a young couple just beginning housekeeping, and trying every possible arrangement of the pictures and bric-a-brac. Even more wonderful is the bower of a bower-bird of New Guinea, — "a cabin in miniature in the midst of a miniature meadow studded with flowers." We cannot forbear quoting from Dr. Beccari's account : —

"This bird," says he, "selects for its hut and garden a spot on the level of the plain, having in the centre a small shrub, with a trunk about

Fig. 464. — Satin bower-birds (*Ptilonorhynchus violaceus*), with their bower and collections.

the height and size of a small walking-stick. Around the base of this central support it constructs, of different mosses, a sort of cone about a span in diameter. This cone of moss seems to strengthen the central pilaster, upon the top of which the whole edifice is sustained. The height of the cabin is at least twenty inches. All around, from the top of the central pilaster, arranged methodically in an inclined position, are the long stems, their upper ends supported on the apex of the pilaster, their lower resting on the ground; and thus all around, excepting immediately in front. In this way the cabin is made, conical in form, and quite regular in the shape the whole presents when the work is completed. Many other stems are then added and interwoven in various ways, so as to make a roof at once strong and impervious to the weather. Between the central pilaster and the insertion in the ground, there is left a circular gallery in the shape of a horse-shoe. The whole structure has a total diameter of about a yard." The straws of which the whole is constructed are the stems of one of the orchids, and they retain their life for a long time after being built into the cabin. Dr. Beccari continues: —

" But the æsthetic tastes of our 'gardener' are not restricted to the construction of a cabin. Its fondness for flowers and for gardens is still more remarkable. Directly in front of the entrance to the cabin is a level place, occupying a superficies about as large as that of the structure itself. It is a miniature meadow of soft moss, transported thither, kept smooth and clean, and free from grass, weeds, stones, and other objects not in harmony with its design. Upon this graceful green carpet are scattered flowers and fruits of different colors, in such a manner that they really present the appearance of an elegant little garden. The greater number of these ornaments appear to be accumulated near the entrance to the cabin. The variety of the objects thus collected is very great, and they are always of brilliant colors. Not only does the bird select its ornaments from among flowers and fruit, but showy fungi and elegantly colored insects are also distributed about the garden and within the galleries of the cabin. When these objects have been exposed so long as to lose their freshness, they are taken from the abode, thrown away, and replaced by others."

First-cousins to these rather plainly colored bower-birds are the celebrated birds of paradise, a group of over thirty species, which rival the humming-birds and the sun-birds in the brilliancy of their plumage. So beautiful are they that we must devote some space to them, and shall largely follow the account of Wallace, the first naturalist to see them in their own forests, using frequently his own words.

When the earliest European voyagers reached the Molluccas in their search for cloves and nutmegs, which were then precious spices, they were presented with the skins of birds so strange and beautiful as to excite the admiration of even those wealth-seeking rovers. The Malays called them God's birds; and the Portuguese, finding that the skins had neither feet nor wings, called them birds of the sun; and the Dutch, birds of paradise. One old author of 1598 tells us that no one had seen these birds alive; for they live in the air, always turning towards the sun, and never lighting on the earth until they die, for they have neither feet nor wings, as can be

BIRDS OF PARADISE.

Great bird of paradise (*Paradisea apoda*), above; six-shafted bird of paradise (*Parotia sefilata*), centre; king bird of paradise (*Cicinnurus regius*), below.

seen from the skins brought to Europe. Even as late as 1760, when Linnæus described the large bird of Paradise, not a perfect skin had been seen in Europe; and that father of Natural History called this species *Paradisea apoda* — the footless paradise-bird.

The birds of paradise are of moderate size, and in structure and habits they are much like the crows and starlings; but they are characterized by extraordinary developments of plumage unequalled in any other group of birds. In some species large tufts of delicate bright-colored feathers spring from each side of the body, beneath the wings, forming trains, fans, or shields; and the middle feathers of the tail are often elongated into wires, twisted into fantastic shapes, or adorned with the most brilliant metallic tints. In another set of species these accessory plumes spring from the head, the back, or the shoulders; while the intensity of color and of metallic lustre displayed by their plumage is not to be equalled by any other birds, except perhaps by the humming-birds, and is not surpassed even by these.

Our plate shows three species of paradise-birds. At the top is the great bird of paradise, the best-known species. It is a large form, and the male alone is thus splendidly ornamented. In the Aru Islands it is very active and vigorous. At one season of the year they have their 'dancing-parties' in certain trees of the forest. A dozen or twenty full-fledged males will congregate. The bird is about the size of a crow, and of a rich coffee-brown. The head and neck are a pure straw-yellow above, and metallic green beneath. At these times of excitement the wings are raised vertically over the back, the head is bent down and stretched out, and the long plumes are raised up and expanded till they form two magnificent golden fans, striped with deep red at the base, and fading off into the pale brown tint of the finely divided and softly waving points. The whole bird is then overshadowed by them; the crouching body, yellow head, and emerald-green throat forming but the foundation and setting to the golden glory which waves above.

At these times the hunters get the birds. They find out what tree the birds have decided upon, and build a little shelter of palm-leaves among the branches. Here before daylight the hunter takes his place, armed with a bow and a number of light blunt-headed arrows. At sunrise the birds begin their dance; the hunter shoots them with the arrow, and they fall to the ground to be picked up by a boy, without a drop of blood soiling their plumage. To prepare them for the market, the body is skinned, the legs and wings cut off, and the skin, stretched on a stick and enveloped in palm-leaves, is dried in the smoke. Did space permit, we would gladly refer to others of this beautiful group, but we can only direct the reader to the chapter on these birds in Wallace's 'Malay Archipelago.' We would remark, however, that while Wallace enumerates but eighteen species, subsequent collectors have increased the number to over thirty.

Our sombre-colored crows, ravens, and the like are closely related to the magnificent birds just alluded to, and here, too, come the series of loud-voiced, scolding jays. Among the birds none make more interesting pets than the crows and ravens. Their voices, to be sure, are not melodious, and their colors are not beautiful, but there is a cunning, a sense of humor among these birds which will afford unceasing amusement. The only drawback is their thievishness; everything bright will be stolen whenever the bird has the slight-

FIG. 463. — Blue-jay (*Cyanocitta cristata*).

est chance. If a thimble be missing, the crow's nest is the first place to look for it; and this search invariably reveals other stolen things concealed among the sticks and straws.

Almost equally interesting are the starlings of the Old World, which we must dismiss with the mention of but two species. First comes the huia-bird of New Zealand, shown in the cut. The most remarkable peculiarity in this species is the difference in the shape of the bill in the two sexes, — short and straight in the male, long and curved in the female. Correlated with this difference in bills is one of habits. The birds are something like woodpeckers in their food, eating the grubs and larvæ in decaying wood. The male uses his sharp bill, like a pick, to tear away the wood, while the female inserts her long bill into the borings, and picks out the morsel which her lord cannot reach.

The other starling-like forms to be mentioned are the buffalo-birds of the tropical portions of the eastern hemisphere. Of these there are several species, but they are all much alike in their habits, and one quotation from the pages of Mr. H. O. Forbes will illustrate the economy of all. "I never tired of watching the friendly relations between the buffalo-birds and their bovine hosts. They used to collect in impatient flocks about the hour of the return of the herd to the feeding-grounds from the wallowing-holes, whither, in the heat of the day, they retired; and as soon as the cattle

arrived, they would alight on their backs in crowds, to the evident satisfaction of the oxen, which they relieved of troublesome parasites. Although
the herd-boys commonly lay dozing at full length on the buffaloes' backs,
the birds seemed to know that they were quite safe, and would even alight
on the bare backs of the sleepers, and from that hop down to the haunches
of the quadrupeds."

Fig. 466. — Huia-bird (*Heteralocha acutirostris*).

The sun-birds are usually confounded in the popular mind with the
humming-birds, although the two are very different in structure and
inhabit different parts of the world. The humming-birds are exclusively
inhabitants of America, while the sun-birds, their rivals in beautiful plumage, belong solely to the eastern hemisphere. They have the same bright
colors, the same metallic lustres, and essentially the same nectar-sucking

habits, occasionally hovering about a flower in the same way, and at other times perching upon a leaf or twig, while with its bill it probes the recesses of a blossom.

The term creepers is well applied to the little climbing birds that bear it. They creep about on the trunks of trees much as do the woodpeckers and nut-hatches, prying into all the cracks in the bark for the insects on

Fig. 467. — Sun-bird (*Nectarinia metallica*).

which they prey. They move about with the utmost ease even when underneath a slightly inclined branch, the stiffened tail-feathers aiding in supporting the body. Allied to these forms are the honey-creepers of tropical America; but while our creepers are sober-colored, the honey-creepers are much brighter. These latter birds suck the honey from the flowers, and, like the humming-birds, feed upon the minute insects attracted to the blossoms by the sweets afforded there. One species of the West Indies

frequently builds its globular nests on the same branches that support the
large structures of the paper-wasps. The reason for this association is
uncertain; it may be that the presence of the wasps is a protection to the
little birds.

Our warblers — to be numbered by dozens — are considerably different
from those of Europe. They are all small birds, whose cheery warble
enlivens our woods and fields all summer long. Some stay in one locality
but a short time, while others make a longer tarry, so that there is scarcely
a moment when one or another cannot be heard. They are exceeding

Fig. 468. — Common brown creeper (*Certhia familiaris*).

busy little bodies. Go into some little grove if you wish to see them and
hear them in all their glory. At first not a note is audible, not a bird
visible. Soon, however, a quick and pretty note is heard from a neighboring shrub, and then comes the song, varying in character with the
species. You sit still, and soon the alarm is over, and the little songster
displays itself in all its naturalness. Now it takes a sudden dart after
some passing fly, and again it is gathering dried grass, or perhaps the
down from yonder dandelion to make its pretty nest. Again it lights and
sings its song, to be answered by some comrade, perhaps its mate, from a
neighboring tree; and so the day goes on, each moment furnishing some
new feature to the careful observer.

While watching in this way there comes a sudden flash of red through the bushes, and one of our most beautiful birds, the scarlet tanager, appears on the scene, lights upon a branch, and shows you plainly his scarlet plumage and his velvety black wings and tail. Notice him well; for he is one of our five representatives of a group numbering nearly three hundred and fifty species, mostly centred in the forests of South America, where forms occur which rival the humming-birds and parrots in their splendor. None of the group are remarkable as songsters, our scarlet species having a song much like that of the robins.

Fig. 460. — Black-throated green warbler (*Dendroeca virens*).

The weaver-birds, two hundred and fifty in number, all belong to the warmer regions of the Old World, the great majority being African. All are remarkable for their peculiar nests of woven straw. Some of these structures are shaped like the retort of the chemist, the entrance being through the long neck. At this portion, which hangs downward, the texture is much weaker, so that any snake or monkey attempting to enter through it will be sure to tumble. Some of these nests are hung from high trees, while others, as in the following quotation from Mr. Forbes, are placed in more humble positions. His description relates to a Javanese species : —

" In a neighboring clump of canes a colony of yellow weaver-birds had thickly hung their nests. Each nest was artfully suspended between the interlacing leaf-stems of one or two reeds in the most skilful way, to secure as much as possible the safety of their eggs, during the waving of the reeds in the wind. These nests were not strung fast to, but strung lightly on, the leaves, sometimes passed through the fork of another leaf, to form a pulley, so as to permit, by sliding along in the swaying of the grass, of their retaining a vertical position, which they must do, weighted as they are by a layer of clay in the bottom of the nests. . . . I was also struck by the fact that different individuals had adopted different forms of nests, which, though agreeing fundamentally, exhibited considerable variation.

The bulk of them were of the retort shape, set with a long-necked orifice hanging downward, but a considerable number, of the progressionist party perhaps, had inaugurated a new fashion by inverting the retort and shortening the neck, giving the doorway an upward and forward entrance, which, if more enticing to depredators, may perhaps be less awkward to the owners."

Still another type must be referred to. In the background of Figure 421 the reader will see a couple of curious roofs in the tree. These are the coverings constructed by the sociable weaver-birds of South Africa, and beneath them the nests of the colony are hung close together, sometimes as many as forty being covered by a common roof. This roof, which is constructed by the whole colony, and which is very heavy and strong, not only protects the nests from the rain, but also prevents any enemy getting access from above. A snake, for instance, might climb the tree and get upon the roof, but he would find difficulty in turning the eaves and reaching the eggs, and not unlikely he would slide from the smooth straw thatch before he got so far.

The American orioles, or hang-birds, build somewhat similar nests, the Baltimore oriole frequently hanging its home on the extremity of the limb of some wide-spreading elm out of the reach of foes of every kind. Here in this purse, half a foot in depth, the flesh-colored eggs are safe, and the callow brood are rocked by every breeze.

FIG. 470.— Baltimore oriole (*Icterus galbula*).

Our next cut shows the great crested cacique of Brazil and its similar but much larger pendulous nests, which may be two feet in depth. Concerning an allied species of the same region, Mr. Bates has the following notice. The Japím " is social in its habits, and builds its nest, like the English rook, on trees in the neighborhood of habitations. But the nests are quite differently constructed, being shaped like purses two feet in length, and suspended from the slender branches all round the tree, some of them very near the ground. The entrance is on the side near the bottom of the nest. The bird is a great favorite with the Brazilians of Pará; it is a noisy, stirring, babbling creature, passing constantly to and fro, chattering to its comrades, and is very ready at imitating other birds, especially the domestic poultry of the vicinity. There was at one time a weekly newspaper published at Pará, called 'The Japim,' the name being chosen, I suppose, on account of the babbling propensities of the bird."

The nesting habit of the cow-bird is much different from that of the hang-birds. This form, like the cuckoo of Europe, is a parasite. It goes around and lays its eggs in the nests of other birds, being especially fond

Fig. 471. — Great crested cacique (*Ostinops citrius*).

of thrusting its progeny upon the care of some of the warblers. While most of the birds are fond of their young, the cow-bird, as one might say, puts hers out to nurse. The young bird, when it hatches, monopolizes the care of its adoptive parents, and quickly thrusts its foster-brothers out of the nest. Sometimes, however, the birds refuse to be imposed upon. When the mother finds the strange egg in her nest, she proceeds to put in a new bottom, completely covering the cow-bird's egg, and thus preventing its hatching.

There is one bird which must not be forgotten. Each spring it leaves the south, and with the successive steps of its migratory progress it leaves a name behind. In Jamaica it bears the name of butter-bird; when it reaches Georgia and the Carolinas, it is the rice-bird. In Maryland and Pennsylvania it takes the name of reed-bird, and then farther north its pretty song gives it the cognomen of bobolink. Everywhere it is persecuted; and its body, merely a mouthful, is the delight of epicures. Reed-birds are now not so abundant as formerly, while the demand for them has increased. This has resulted in one great good. To supply the market, that unmitigated nuisance the English sparrow is killed, and after the beak is injured and the bird is deprived of its feathers, it takes a refined taste to tell the difference between the two.

FIG. 472. — Bobolink (*Dolichonyx oryzivorus*).

The last family of the birds is that which contains the finches. It is very large, embracing as it does some five hundred species. Here come the grosbeaks and the cross-bills.—the one with large and stout bills suitable for cracking seeds and nuts; the other with the bill adapted to extracting the seeds from the cones of the various pines and firs. Once in a while a cold winter will bring the pine-grosbeak down from the north into New York and New England in large numbers. Living as they ordinarily do in the far north, away from the haunts of man, they are remarkably tame, or rather unsophisticated, when they come among us. No birds are easier to capture than they. A noose on the end of a walking-stick is all that is necessary, and one can readily approach them, and put the noose over the head of one without alarming the others of the flock. More common are the rose-breasted grosbeak and the cardinal-bird represented in our cut. Here, too, must be mentioned the indigo-bird, the song-sparrow, the chipping sparrow, the snow-bird, the canary, and a host of others, one of which needs a paragraph by itself.

Some years ago some persons not overblessed with brains introduced into the parks of New York the English sparrow. It was confidently expected that this bird on entering our country would change its habits and here feed upon worms. In only one way did it fulfil the expectations

of its admirers: it increased with great rapidity, and now the quarrelsome, useless creatures are everywhere to be found. They care nothing
for worms, and are not of the slightest use to the agriculturist. They
live on the crumbs and grain thrown out by charitable persons, and on

Fig. 473. — Cardinal grosbeak (*Cardinalis cardinalis*), above; rose-breasted grosbeak (*Habia ludoviciana*), below.

what they can scrape from the dropping of horses. They let the canker-
worm, the potato-beetle, and the cabbage pests alone. They are quarrelsome, and tend to drive away our much better native birds. Fortunately
they are good to eat, and the sooner the law withdraws its protection from
them, and the sooner they appear on our bills of fare, so much the better
for the country.

MAMMALS.

The mammals of to-day are as sharply marked off from all other animals as are the birds. With the exception of the whales and some other fish-like aquatic forms, there never has been the slightest doubt concerning a single member of the group. The most superficial character, the hair, is at once distinctive; for every mammal has hair, and nowhere else in the whole animal kingdom do we meet with a similar protection to the body. The scales of fishes and reptiles, the feathers of birds, are totally different in nature and appearance. Even the whales have hair in their young stages, and in some it persists even in the adults.

Though lacking the bright plumage of the birds, the mammals have their own peculiar points of interest in which they are not excelled by any other animals. First, their size varies between the greatest limits; the little harvest-mouse of Europe weighs scarcely more than an old-fashioned copper cent, and between this and the monstrous elephants and whales almost every grade can be found. In shape, too, there are many interesting points. The long-necked giraffe, the almost neckless fish-like whales, the armored armadillos, the bird-like bats, and the almost human apes need only be mentioned. So, too, the mental side has its interests; for here we meet with every degree of intelligence. Some are models of stupidity, and, on the other hand, man is a mammal, and in him the intellect reaches its highest development.

THE DUCK-BILL AND THE ECHIDNAS.

Away down at the base of the mammals are three or four peculiar forms which to the naturalist are highly interesting, from their strange structure and development. All live in that land of curious animals, Australia and Tasmania, some stretching over into New Guinea. Their greatest peculiarity lies in their resemblance to the birds and reptiles in many points of anatomy and development, while, on the other hand, they are distinctly mammalian in the rest of their anatomy.

First, and strangest, is the duck-bill, an animal about a foot and a half in length, covered with a dark brown fur, its feet webbed, its tail shaped much like that of a beaver, while its head terminates not in toothed jaws like an ordinary mammal, but with a bill much like that of a duck, three inches long and two in breadth. At the base of the bill is a fold of skin which is capable of being thrown back over the eyes. If we look at the inside of the beak, we see that its resemblance is even closer to that of a duck than it appeared from the surface. As will be remembered, in the ducks there were ridges which served to strain the water from the food, and in this animal exactly the same structure is seen — four horny teeth, which evidently subserve a similar purpose; true teeth occurring only in early life.

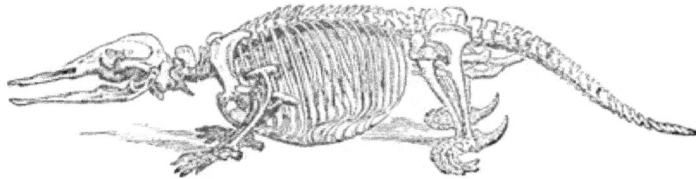

Fig. 474. — Skeleton of the duck-bill.

The duck-mole (another of its names) lives an aquatic and a burrowing life. A little above the water are seen the entrances of the serpentine burrows, which go obliquely upwards and terminate in a large chamber, where the nest of grass and dried leaves is placed. In hunting for their food they paddle along, with the head beneath the water like a duck, turning over the sand and stones, and sifting out the insects and crustaceans which abound in such localities. These are stored in pouches in the cheeks, and then, after the fishing is over, they are crushed and ground between the horny teeth. The web of the foot, so useful in swimming, would be in the way while digging or traveling through the burrows, and so it folds over on the sole of the foot when the animals leave the water.

Fig. 475. — The duck-bill (*Ornithorhynchus paradoxus*).

The echidnas, or porcupine ant-eaters, are much different in their external appearance. The duck-like bill and webbed feet are lacking, while the mouth is but a small orifice at the end of the snout, merely large enough to allow the tongue to pass in and out. Then, too, the fur is different; for among the hair are spines two or three inches long, which are much like those of the more familiar porcupines. As this difference in structure would indicate, there is a great difference in their habits; for the latter animal is a terrestrial species, which burrows in the earth in the search for ants, which form its food. When a nest is found, the animal sticks out its long tongue covered with an adhesive saliva, and then draws it in covered with ants.

There are some two or three species of ant-eaters, one occurring in Australia and reaching over into Tasmania, while New Guinea apparently has two species. All are small, about a foot in length, and the females have, during the breeding, a rudimentary pouch for carrying the young, which, as it were, foreshadows that of the kangaroos and opossums soon to be described.

Technical matters are not admitted to our pages, and so we must forego any account of the peculiarities which ally the duck-bill and the spring ant-eater to the reptiles and birds. One point is, however, so interesting that it must be mentioned. In the early days of scientific natural history it was universally believed that these animals laid eggs, and the eggs were even figured. As time passed, this idea fell into discredit, the eggs were regarded as those of some reptile, and at last the whole thing was almost forgotten, and these animals were universally regarded as bringing forth their young alive, as do all other mammals.

Such was the state of affairs until the last half of the year 1884, and then by one of those remarkable coincidences the true facts of the case were discovered almost simultaneously by two naturalists hundreds of miles apart. On the second day of September, 1884, the British Association for the Advancement of Science, then holding a meeting in Montreal, were startled by a cablegram of four words from Mr. Caldwell, then in Australia, announcing that these animals were oviparous. The telegram was sent to London, and thence was repeated to Australia, and appeared in the 'South Australian Register' of the fifth, and in the same number was an account of a scientific meeting held in Adelaide on the second, in which Dr. Haacke announced the same discovery. The coincidence was even more close; for Dr. Haacke made his discovery on the 25th of August, while Mr. Caldwell, two days' journey in the bush of Queensland, started his message at almost exactly the same time. The egg is about the size of that of a crow, and is covered with a flexible white shell, and the young duck-bill is provided with a little knob on the beak, like that which aids the chick in escaping from the egg.

POUCHED ANIMALS.

We have just alluded to the little rudimentary pouch of the spring ant-eater, in which the young are carried for a time, but in a large number of otherwise less anomalous mammals the same structure acquires a far greater development. In fact, it forms a regular pocket on the belly, in which the young can stay until they reach a size when they are able to take care of themselves; and from this pouch these animals receive their name Marsupials.

First, before describing any of these anatomical peculiarities, let us consider the distribution of these forms. In America occur the opossums, but, with the exceptions to be noted immediately, nowhere else in the world can you find any more such forms. In Australia, on the other hand, these forms abound, while a few stretch south into the adjacent island of Tasmania, and north into New Guinea and Celebes and a few neighboring islands. Australia, however, is the great centre, and before the arrival of the whites it contained no other terrestrial mammals, with the solitary exception of the dingo, or native dog, which may have been introduced by man.

Another equally interesting feature is the way in which many of these forms simulate other groups of mammals. Thus we have forms like bears, squirrels, dogs, rats, gophers, and the like; and if we take the fossil forms into consideration, the lions, too, have their representatives among these curious animals. This resemblance extends not merely to the external form, but affects as well the character of the teeth and of the digestive tract. The reason for this is not far to seek. In the region occupied by those animals all activities of life were open without competition, and one form adopted one habit, and another another, and with it came a change of structure better adapted for the mode of life. Formerly habits were regarded as the results of structure, and many pages have been written to show how every feature of anatomy was adapted to the mode of life. The inference was that forms were created thus, and that habits were the result of structure. Now we know better, and the converse of the old idea is now admitted by all to be true, — structure is secondary, and is produced and modified by habits.

First in order come the American opossums, typified by the common opossum of the warmer parts of the United States, which is well shown in the accompanying cut. It is much like a rat in its appearance, except in size. It has the same pointed snout, the same pointed ears, and the same round scaly tail. This tail, however, has a function unknown to that of the rat; for it is prehensile like that of some of the monkeys. It can be curled around the limb of a tree with sufficient strength to support the body. The opossum is largely a nocturnal animal. In the day-time it lives in some hollow tree; but at night it descends, to begin its wanderings and its search after food, largely of insects, varied with an occasional reptile or bird, and, when convenient, with eggs. The Texan variety is said to be very fond of the black persimmon.

The common opossum has a well-developed pouch in which it carries its young. For the first two weeks the mother keeps the opening of the pouch closed; but after that time it is opened, and when the young (from nine to thirteen in number) are about four weeks old, they occasionally leave the pouch and climb upon the mother's back; and then when Mrs. Opossum takes her family out for a ride, the scene is one which is not readily duplicated. The little ones are nestled in the dirty yellow fur, and each has its tail elevated in the air and twined around the caudal appendage of the mother. In confinement, the opossum has but few interesting features, and displays but little of those intelligent actions which make so many other animals interesting. This is what we should expect from the small size and general character of its brain. The economical value of the opossum is but slight: the negroes are fond of its flesh, but whites do

FIG. 476. — Virginian opossum (*Didelphys virginianus*).

not find it over-palatable. There is, however, one side to this animal which must not be omitted; for it has furnished us with the expression 'playing possum.' When this animal finds that all attempts at escape are useless, it feigns death in the most perfect manner. No amount of tormenting will make it show the least sign of life. It will remain in any position, no matter how uncomfortable; it will stand any amount of pinching and poking without the slightest movement of a muscle, awaiting the time when its tormentor will leave it for dead.

In tropical America there are many other opossums, some with well-developed pouch, others with it in a very rudimentary condition. Some live on insects, some on reptiles, others on the banana and other fruits, while still others turn to the water for their means of subsistence. These latter have the feet webbed much like those of a duck, and they swim and dive with the greatest ease. Their food is largely fishes and crabs. In one — the water-opossum — this aquatic life has gone so far that the animal has much the appearance of an otter, thus affording one of those interesting resemblances referred to above.

All of the remaining pouched animals are confined to Australia, New Guinea, and the adjacent islands. Some feed solely on vegetable substances, and some are as thoroughly carnivorous as any wolf or lion, and still others have a diet of insects.

First comes the wombat, like the prairie-dog in its burrowing habits, and in the shape and structure of its skull. We shall shortly see how the squirrels and their allies have the front teeth like chisels, the loss by wear being constantly made good by a continual growth at the roots. The wombats, like these forms, are gnawing animals, and here the same structure of the teeth is produced. The koala, or Australian ‘bear,’ on the other hand, has a bear-like look, and at the same time has many of the features of the sloths. It is thoroughly arboreal in its habits, climbing the trees with the greatest ease, but when on the ground moving about with considerable awkwardness and difficulty. A glance at the feet explains this. They are armed with long, curved claws, and in the fore feet the toes, something like those of the chameleon described on page 406, are divided into two bunches, two toes being opposable to the other three in much the same way that our thumb is related to our fingers.

The various species of cuscus most nearly resemble the opossums of any of the eastern forms. They have the same prehensile tail, the terminal half of which is bare. They also live in trees, moving about slowly and feeding upon the leaves, of which they devour large quantities. They possess great vitality, and are not easily killed with a gun. A charge of shot frequently will be stopped by the thick fur, and even breaking the spine or piercing the brain will not kill them for some hours. They are hunted by the natives for food. Their slow motions render it an easy task to capture them by climbing. Their fur, too, is highly esteemed, and is used in the manufacture of ornaments for the person — clothing is scarcely necessary for the natives of these warmer climes. A relative of the cuscus — the sugar-squirrel — is noticeable from the fact that it has the same habits as our little flying-squirrel. Like it, it lives in the trees, and it has the same web of skin between the legs, and the same flattened tail. Like the flying-squirrel, it takes long, sailing leaps from the trees. There are also larger forms which are more like the taguan, to be mentioned farther on. Still other forms must be mentioned from their resemblance to dormice.

Possibly the most celebrated of all these animals are the kangaroos, the largest species of which is shown in our plate. Here the comparisons which we have noticed in other forms are largely lacking. The kangaroos have no close parallel in the other groups, unless we consider the jerboas and the jumping-mice in this connection. In the typical forms there is a great disproportion between the fore and hind legs, the former being small, the latter enormously developed. Then, too, the tail is very large, and to a certain extent serves as a fifth leg. When the animal is at rest, it sits, as it were, on a three-legged stool, its hind legs and its tail all entering into the support; but when in rapid motion, the tail is stretched out straight, the fore legs are closed against the breast, while the hind legs are the sole organs of locomotion. The ordinary leap is nine or ten feet, but when alarmed this may be even doubled.

It is an easy matter to frighten these animals; any strange object will produce an alarm, and specimens in captivity have even been frightened to death. If cornered, however, it is not a mean antagonist. It grasps its opponent with the short front legs, and then kicks, or rather claws, with the hinder ones, the sharp nails of which tear clothing, flesh, and everything. Kangaroo-hunting is a favorite sport in Australia, and the game is taken in many ways. The kangaroos are grass-feeders, and their jaws and teeth are adapted to cropping the pasture very close, and thus they can subsist in a region where any ordinary form would quickly starve. This same feature makes them a nuisance, for they cut off the grass so closely that they kill the pasture, and this fact has led to their extermination in the more thickly settled portions of the island-continent.

In size and appearance there is considerable range in the kangaroos. Some are adapted to live on the open plains, while others are fitted for rocky regions, and strangest of all are the tree-kangaroos of New Guinea.

There is a great tendency on the part of a certain class of people to apply the name of his Satanic majesty to almost every striking object of nature; 'devil's dens,' 'devil's punch-bowls,' and the like abound. But while we may not appreciate the taste which leads to such adjectives, we must admit that in one instance the term was well applied. The Tasmanian devil, both in appearance and character, deserves the name. It lives, as its name indicates, in Tasmania. It is fond of flesh of all sorts, and formerly did considerable damage to live-stock and poultry. Now it is only found in the wilder portions of the island. In captivity it is always the same snarling, growling beast, fighting with its companions, and never showing the slightest disposition to become tame. During the day it tries to hide in the corner of its cage, and at night only is it active.

LARGE KANGAROO (*Macropus major*).

A relative is the zebra-wolf, or pouched dog, of the same island. Its name indicates its appearance; a shape like a dog or a wolf, and stripes of black across the grayish brown fur of the back. Like the devil, it is carnivorous, and formerly it was a great enemy to the sheep and poultry. This led, however, to a war of extermination on the part of the colonists, and now the pouched dogs are rare. Another form to be mentioned is the banded ant-eater of western Australia, which here takes the place of the true ant-eaters of the tropics. It has strong paws adapted to tearing open the hills of the ants, while its tongue is long and slender, and covered with a sticky saliva.

Fig. 477. — Tasmanian devil (*Dasyurus ursinus*).

We must not dismiss the marsupials without a reference to some of the features of their reproduction. The young of all are little, blind creatures, the largest not over an inch and a quarter in length. As soon as they are born, the mother takes them one by one in her mouth, and places them in the pouch, with the mouth upon the teat. The creature is too small to suck, and so the mother has the means of forcing the milk out into the mouth of the young. Were this the case with other mammals, the young would certainly be choked when trying to breathe; but here there is a provision against such a danger, for the windpipe, instead of terminating at the throat, as in other forms, is here continued up into the back part of the cavity of the nose. In the case of the large kangaroo the young are suckled for eight months, and even after they are able to run about and browse upon the grass, they flee to the pouch on the slightest alarm, even until they reach a weight of ten pounds.

EDENTATES.

The term edentate means toothless; and in applying it to the present group of animals the reader must not understand that the members are without teeth, but merely that they lack the front, or incisor, teeth. There are other mammals with the same feature, but various anatomical charac-

FIG. 478.—Three-banded armadillo (*Tolypeutes tricinctus*).

ters (too technical in nature to be discussed here) serve to mark off the present forms from all the rest.

The curious armadillos of tropical America are possibly as familiar as any member of the group, and in this connection we may mention the fact that these forms are far from being literally toothless, for some of them have an abundant supply, one species having a full hundred in both jaws. The armadillos derive their name from the armor enveloping the body. This armor is formed by the deposition of bone in the integument of the body, and it may cover not only the back, but the top of the head and tail as well. In most forms this coat of mail is jointed like that of the ancient knight, and the various portions are movable upon each other, thus allowing the animal to roll itself up in a ball, and leaving no vulnerable portion exposed. Our cut shows two individuals thus curled up. In this species, as will be seen, there are four of these joints, and between them are three belts of bony plates which give the animal its name. In others the number of these bands may reach as many as nine, as in the case of the only species which enters the United States.

The armadillos are all burrowing animals, digging their holes some six or seven feet in depth by means of their strong, long-clawed feet. In these burrows they spend most of the day, and at night they come out to seek their food. They are rather indiscriminate in their diet, eating either vegetables or meat with equal relish, and some of the larger forms are said to occasionally burrow into graveyards. Yet notwithstanding their omnivorous character, these animals are highly esteemed as food. When caught above ground they quickly roll themselves into a ball, but the defence which answers so well with other animals is no trouble to man; he simply takes the animal, shell and all, and then when he gets it home, this same shell serves for a dripping-pan, for the animal is roasted in it.

The strangest-looking of all the armadillos is the pichiciago of Chili. It is only about six inches in length, and has the plate armor only upon the head and back, leaving the sides exposed, except so far as they are covered with hair. The hinder part of the body is the strangest; for it looks as if it were chopped off square, and the little armored tail comes out in the most curious manner.

While the armadillos are in every way fitted for a burrowing life, another group of the same edentates are pre-eminently adapted for a life in the trees. These are the sloths; and they are rightly named, for their slowness is proverbial. They move about the branches of the trees, hanging down from the limbs by means of their long claws. Here they live and sleep and eat, and but rarely, if ever, do they descend to the earth of their own accord. As in action so in intellect they are slow. Linné was right when he applied the term Bruta to these and other allied animals, for they are the most brutish of all the brutes. A sloth, says Dr. Oswald, "permits you to lift his claw, but drops it as soon as you withdraw your hand. If you prod him, he breaks forth into a moan that seems to express a lament over the painfulness of earthly affairs in general rather than resentment of your particular act. . . . I do not know if a sloth can be teased; for after trying all my conscience and Mr. Bergh would permit, that point still remains undecided."

They are, farther, slow to die. If one be shot in the top of some tall tree, it comes crashing down through the branches, grasping frantically for some support, and the chances are that its fall will be stopped ere it reaches the earth, and then if death ensue, the long claws will not let go their hold. Poisons, too, are slow in taking effect on them. But slow as they are in these respects, they are seen at their worst when placed on level ground. Their feet are like pot-hooks, only adapted for hanging to

FIG. 479. — Three-toed sloth (*Bradypus tridactylus*).

some branch, and when on the ground they are almost perfectly useless. If there be some stump or stone, they may pull themselves along by its aid, but without such they are almost perfectly helpless. And yet they are able to swim, and to make pretty good progress; for Wallace and Bates saw one in the region of the Amazon, swimming a river where it was a thousand feet broad.

There are two distinct types of sloth; one with two, the other with three toes to the feet. There are some twelve species known in all, some ranging north into southern Mexico, but most of them living in the dense forests of the Amazon and the Orinoco. They have but little that is of interest in their appearance; in museum specimens the hair of the body is some shade of a dirty gray, but in life it has a decidedly greenish hue, the cause of which is interesting. It is caused by the growth of a minute plant, allied to the fresh-water algæ, upon the hair.

FIG. 480. — Great ant-eater (*Myrmecophaga jubata*).

The true ant-eaters have much the same range as the sloths, and are only excelled by them in slowness and awkwardness of motions. Our cut shows the largest species, and some comments upon it will serve to illustrate some of the features of these animals. In the first place, the figure flatters the animal; it makes it much too good-looking. In nature the hairs of the enormous tail are stiff and like wires, and nowhere does the pelage have that softness which one would expect from the figure. One of the fore paws is shown with its huge claws much like those of the sloths. These are used to scratch open the ant-hills so abundant in tropical America, and then when the hill is open, that long tongue shown coiled like a string on the ground is protruded from the long, tubular muzzle, thickly coated with a sticky saliva, and then, covered with ants, is drawn into the mouth. The structure of the fore feet makes these animals very awkward upon the ground, and as they walk, the toes are doubled in under, and the functional sole is formed by the side of the foot. Others of the group are almost as arboreal as the sloths, seeking their insect prey in the trees.

The aard-vark of the Booers of South Africa (earth-hog the name means) combines the habits of both the ant-eater and the armadillo, and has but little of the hog in its appearance. Like the armadillo it lives in burrows beneath the ground, and like the ant-eater it is insectivorous in its habits. It walks better than its relatives, and when night begins, it sets out on its nocturnal rambles. With its short, broad claws it digs most rapidly, and by means of these it is able to quickly make a hole in the hard domes of the termites. Its tongue is not so long as that of the ant-eater, but it is covered with the same viscid saliva, and is used in the same way in taking the food.

The pangolins of Asia and Africa are armored forms, but their armor differs from that of an armadillo, being composed of large scales arranged like those of a pine-cone. Their habits need a word. They, like others of their relatives, are ant-eaters, and their large claws are used in the same way in digging, not only the holes in which they live, but into the nests of the ants and termites. Some of these are able to climb trees by means of the claws and the tail, which is somewhat prehensile. In confinement, according to Sir Emerson Tennant, in his charming volumes on Ceylon, they make very gentle and affectionate pets, and in nowise seem to deserve the name Manis — a ghost or departed spirit — which Linnæus gave them.

RODENTS.

The rodents are gnawing animals, familiar to all in rats, mice, squirrels, and the like. In number of species they far excel all the other groups of mammals, and the same is true of the individuals. The swarms of rats of our cellars and barns, sewers and wharves, are beyond enumeration; but even these do not compare with the mighty hordes of lemmings which live in the Arctic regions. And yet all this series is of comparatively little use to man. Some afford him furs and skins, and a few are of use as food, but the great majority are almost unmitigated nuisances.

There is one structural feature to which we would call attention before the enumeration of the species. This is the structure of the front, or incisor. teeth. If we examine the jaws of a rat. we find that the two front teeth in either jaw are always sharp and chisel-shaped. This is the result of their structure. The hard enamel is all on the outer side of the tooth. while the inner portion is composed of softer dentine; so in use the softer portion is worn away much faster than the hard, and in this way the cutting edge is always kept sharp. Were these teeth like those of man, and exposed to the hard usage of those of a squirrel. or a beaver, they would soon become worn down to the gums. But in the rodents this does not occur: the tooth is constantly growing at the base, and thus the loss is continually made good. It occasionally happens that this growth does not work for the benefit of the individual, no matter how good it is for the race. If one of these teeth be broken or destroyed. the opposing tooth is not worn away as it should be, but its growth continues just the same. and so the tooth. curving as it grows. soon passes the mouth. and if it be in the lower jaw. it may eventually grow around until it pierces the skull. utterly barring the mouth, and eventually causing the death of the unfortunate by starvation. even in the midst of plenty. Making the proper changes in the words. the same account will apply to the teeth of the upper jaw.

The first of the rodents to be mentioned are the well-known hares and rabbits. long-eared, jumping animals, mostly confined to the northern hemisphere. They are far from courageous animals; they are rather the personification of timidity: they have no cunning and no means of defence; when danger comes, they must take to their legs or to their burrows. The differences between hares and rabbits are very slight. In strict propriety. the latter name is applicable only to the species of southern Europe. which. in a state of domestication. has been distributed all over the world: all the rest of the long series are hares. There is no necessity to describe all these; but a few must be mentioned at least by name. Among these are the long-eared. long-legged jackass-rabbits of the plains. whose long leaps and rapid gait are referred to in every book of travel. Then. too. the 'Mollie cotton-tail' of the south, the wood-hare of the north. which nowhere seems to have even a fraction of the cunning and address of the 'Brer Rabbit' of Uncle Remus.

There is one aspect of rabbit-life which is interesting. Some of the northern species are dark-colored in summer and white in winter. The extent to which this process is carried varies considerably. In some the sides and limbs retain a tawny shade; while in others the whole animal becomes a snowy white, excepting the tips of the ears. which remain black. These hares which change in this way are called varying hares. and this change in color is of the same nature as that found in the ptarmigan, the

Arctic foxes, and the like. It is protective in its character: for the white tends to make the animal less easily seen on the fields of snow which linger for many months in winter, while the darker color agrees better with the general character of the same region in the short summer when the snow is gone.

FIG. 481. — Hares (*Lepus timidus*).

The pikas, or 'little chief' hares of the mountains of the west, are curious relatives of the true hares. They lack the long ears of the true hares, and even bunny's short tail is lacking. They live among the rocks, in small communities, and lay up a store of provisions in summer to carry them through the winter. They are timid little creatures; but may often be seen sitting upright like marmots on the rocks around their burrows, and uttering at short intervals their sharp, shrill, barking cry.

The largest of the rodents is the water-cavy, or capybara of South America, a huge, stupid beast four feet long, and no more interesting than its relative, the familiar little Guinea pig, which comes from the same region. This latter piebald creature is often kept as a pet; but why, it is impossible to see, unless because it is so easily reared; for in the whole range of mammals there is not a single species which can excel it in stupidity. It knows absolutely nothing, and can learn nothing. The same object that frightens it to-day will alarm it every day for a year. Besides its stupidity it has only one other remarkable feature, and that is the rapidity with which it reproduces. It is uncertain from what wild stock the Guinea pig has descended. There are several wild species, but none of them are marked with the same colors. It seems probable that this species had been domesticated for many years before America was discovered. It may be remarked that the name Guinea pig arose from a confusion of Guinea with Guiana.

The pacas and agoutis are also South American animals with but little of interest pertaining to them. The chinchilla, on the other hand, has a celebrity on account of its beautiful fur, and hence may demand a little more space. They are small animals, only about nine or ten inches in length, and are found only on the western slope of the Andes, living among the rocks. The fur is very thick and soft, and is beautifully marbled with gray and dusky, features which have brought it into great repute. Still, as the skins are small, it takes many to make a garment, and for this reason a chinchilla sack is very expensive, though, of course, the value varies with the caprices of fashion. In shape the chinchillas are something like squirrels, but they lack the bushy tail of these forms. They are active and agile, and in their motions they are much like our familiar little ground-squirrels. Related to them is the viscacha or biscacha of South America, a burrowing form whose holes are found all across the southern part of that continent. There they replace the prairie-dog of our plains, and in their burrows occurs the same species of burrowing-owl as is found in the villages of the prairie-dogs of the north, the association being of much the same nature.

Everywhere one runs across the belief that the porcupines have the power of shooting their quills to a considerable distance at any enemy. The story has been refuted times innumerable, but still the belief remains. No amount of contradiction seems able to put it down. The truth is that these animals have some of the hairs modified into sharp, strong spines, which are easily detached from the skin. The size of these spines varies greatly in the different species. In our common form they are small and but an inch or two in length, but in European and African species they may be a foot in length, and as large in diameter as a lead-pencil. These quills, large and small, are borne on the head, back, and tail; and when the animal is surprised, it does not attempt to run, but simply rolls itself into a ball, with the head between the legs, and the spines standing out in every direction. It is then a ticklish object to handle. In our species the quills, though short, are exceedingly sharp, and each is finely barbed at the tip, so that it is not readily withdrawn from the skin or clothing; but, on the other hand, every motion tends to force it farther in. If a man or a dog approach one of these balls too closely, they are apt to rue their venturesomeness. There is a quick, vicious flirt of the tail, and some of the

Fig. 482. — American porcupine (*Erythizon dorsatus*).

spines are pretty sure to be left in the face of the one, or the hands or legs of the other. This jerking, quick as a flash, is the source of all the stories of shooting the quills.

Besides these differences in the spines of the European and the American porcupines there is one of habit. Those of the Old World live in holes in the earth; those of the New (especially the species figured) are arboreal in their habits, climbing trees with the greatest ease, but making but slow and awkward progress on the ground. These forms, when abundant, may do considerable damage in the forest, as they feed upon the barks of trees. The South American species are even better adapted for a life in the trees than is our form, for they have the tail prehensile.

One of the porcupine-like forms — like the porcupine in structure, but not in appearance — is the celebrated coypu of South America, which furnishes the nutria fur of commerce. The coypu is about the size of a beaver; it is like an otter in its habits, living on the banks of rivers, and swimming by means of its webbed feet, with the greatest ease. Its fur is brown and soft, and as these animals are very abundant, and easily obtained, they have tended to reduce the price and demand for the beaver-skins of North America.

The jerboas of the warmer parts of the Old World are the kangaroos of the rodents. As in those animals, the fore legs are short and weak, while the hind ones are enormously developed, and the animal but rarely attempts any progress except that of jumping. Its leaps are enormous in proportion to the size of the animal; it merely touches the tips of the toes to the ground, and with its long, tufted tail streaming behind, it seems rather to fly than to move by means of its feet. In our jumping-mouse these leaping powers are not so well developed, but still they are considerable. When moving about in the ordinary manner the animal uses the fore feet as well as the hind legs; but when alarmed, only the latter are called into action, and the animal can then jump six or eight feet at a bound. The late Professor Tenney gave some account of this species, and especially its hibernating habits, an abstract of which may prove interesting. The animal was found in January, coiled up in its nest two feet below the surface of the earth, and to all appearance dead, except that possibly it was not as rigid as a dead mouse would be in winter. It was taken home, and after being held in the hand for some time it began to show signs of coming to, and in the same afternoon it became perfectly active. It was given food and materials for making a nest, and as soon as the lights were put out it began its gnawing, and by morning it had a ball of paper and cotton about six inches in diameter, and was snugly ensconsed in the middle of it. The next night was cold, and in the morning the mouse was seen on the outside of the nest apparently dead. So it went on all winter; cold making it torpid, and warmth reviving it and restoring it to a state of activity.

The pocket-mice are somewhat similar to the jumping-mouse in appearance and habits, and form a transition between them and the gophers. Some of them jump as well as the form just mentioned, and all are characterized by having pouches in the cheeks. These pouches, or pockets, do not communicate with the mouth, but open at the sides, and are lined like the external skin with fur. They are somewhat nocturnal in their habits, and while they are frequently seen by day, it is only after dark that they display their greatest activity. They are all confined to the district west of the Mississippi, and some of the species show a distinct tendency to take the place of our common rats and mice about the barns and storehouses.

In the pouched gophers the cheek-pouches acquire their maximum development. They are there, comparatively speaking, huge sacks, extending back from the sides of the mouth to the shoulders; but what use they are to the animal is as yet not thoroughly settled. By some they are regarded as sacks to contain a temporary supply of food, while others think them of use in carrying away the dirt which is excavated in making the burrows.

FIG. 483. — Pouched gopher (*Geomys bursarius*).

This burrowing habit makes the gophers an intolerable pest in the west; for they choose the richest soil for their habitations, and completely honeycomb it with their winding ways. Everywhere their holes may be seen, and scattered here and there are the little heaps of earth brought up from below. The soil is frequently so thoroughly undermined that it slumps through in walking upon it. Besides this burrowing, which renders them a nuisance, they are fond of vegetable matter of all sorts, and especially of bulbous roots, of which they gather large quantities and store them away underground. Every conceivable method is taken to check them, — poison, traps, drowning-out, and the like, are all employed, — but without

causing any great diminution in their numbers. Plugging up their holes
is of no use, for they quickly dig others.

We have various species of these gophers, but all are essentially the
same in habits. The species figured is the best known in the Mississippi
valley; in our southern states, where the term 'gopher' is applied, as we
have seen, to a turtle, the gopher is called a 'salamander'; why, is a
question more easy to ask than to answer. Other species occur on the
Pacific coast, and are there as great pests as are their relatives farther east.

In South Africa, about the Cape of Good Hope, occurs the mole-rat. It
is rat-like in shape and about ten inches long, but has a short tail. It
occurs in immense numbers on the sandy beaches, excavating its tunnels
in every direction. In some places these forms are so abundant, and their
galleries are so numerous, that it is dangerous to ride on horseback along
the shore, so great are the chances of being thrown over the horse's head
by the unexpected sinking of the horse's feet into the holes, or by the
caving in of the burrows.

The family which contains the true rats and their relatives is very
large, some three hundred species belonging to it; but of these only a few
need mentioning. One of the most important members of the group is
the muskrat of North America; indeed, it is almost the only one that has
any great importance. In science it has been very fortunate, for it has
never been the recipient of many names. Of common names it has many
more, — musquash, ondatra, watsuss, wachusk, Massascus, and musk-
beaver. As this last name indicates, the animal shares to a certain extent
the habits of the beaver. This animal is so familiar that it needs but few
words.

The muskrat is largely nocturnal in its habits. It prefers the marshy
banks of rivers and ponds to make its home. It makes its summer resi-
dence in the banks; but for winter it builds a hut in the water much like
that of the beaver, the entrance being below the surface. These huts are
constructed of twigs and mud, and may be from two to four feet in height.
Inside is a dry, warm nest lined with grass. In the winter they do not
become torpid, but are almost as lively as in summer. They leave their
houses, and make their journeys under the ice; but as they are obliged to
breathe at intervals, they have their breathing-spots. These are covered
over with mud on the sides, and have a lot of loose grass in the centre,
which prevents the water from freezing except in the very coldest weather.
Muskrats are trapped in immense numbers for their skins, by means of the
ordinary steel trap baited with sweet apple, or some other favorite food.

The lemmings are inhabitants of the regions around the north pole.
They occur in North America and Greenland, but the species of north-

ern Europe is far more celebrated. It lives in the far north, but at intervals immense hordes, almost as innumerable as the sands of the sea, press southward and westward from the mountains to the plains. These armies follow a straight course; nothing can stop them; they swim the rivers and lakes, and rush through the towns in their way, until at last they are swallowed up in the sea. While on these migrations they suffer from foes of every sort; carnivorous quadrupeds of all sorts grow fat on the bounteous supply, while hawks and owls, swooping down upon them, carry off thousands. Still they press on, utterly oblivious of danger, and apparently of the fate that awaits them. What is the cause of these migrations, seemingly of so little advantage to the race, has not been satisfactorily settled. The best explanation is that they have increased so in numbers in their mountain strongholds that the region is overstocked, and the surplus are forced by circumstances to emigrate, and to try to find a new home elsewhere, an attempt which ends disastrously for the emigrants, but relieves the over-crowded mountains.

The rats and mice proper — those domestic pests which have followed man all over the world — are too well known to call for extensive description, and yet there are some facts connected with them that may bear repetition. First as to the rats. The original house-rat of Europe was the black rat, and from time immemorial until a century ago it swarmed everywhere. It was introduced into America about the middle of the sixteenth century, and multiplied so rapidly that soon this country was as well supplied as the Old World. A change was to occur. As central Asia has been the starting-point for successive waves of man, one after another overrunning Europe, so was it with the rats. From this region the brown rat started. In 1737 it crossed the Volga in immense numbers, overran all Russia, and rapidly spread over Europe. At about the beginning of the Revolution it was brought by ships to our continent. This newcomer was from the standpoint of rats a far better animal than the black rat, and the result was that soon the latter had to give way before his stronger rival. To-day the black rat is extremely rare, but everywhere one finds the brown, or wharf rat, which also bears another name. — Norway rat, — given under some misapprehension as to its origin. The whole of this history presents a curious parallel with that of the human population of Europe and America, not only in the place of origin, but in the pre-eminence taken by the later invaders. The Teutonic has largely supplanted the Hellenic-Latin wave.

This brown rat is a formidable animal. It stands an easy first among all the rodents from its courage, its voracity, and its ferocity. It will eat almost anything, and it attacks and kills animals even larger than itself.

It abounds everywhere, — around the wharves of our seaport towns and in the sewers of the cities; in our houses and in the barns of the country. In comparison the small and really pretty house-mouse, also of Europeo-Asiatic origin, is of no account.

We have a large number of native mice, some much like our common domestic mouse, while others show tendencies towards the muskrats and lemmings. The latter include a large number of field-mice, or voles, while among the former are the cotton-rat, the Florida wood-rat, and the pretty little white-footed mouse of our fields. The cotton-rat and the

Fig. 484. — Mouse (*Mus musculus*).

Florida wood-rat are, as their names would indicate, inhabitants of the southern states. The latter species lives in the woods and makes immense nests of twigs and dried leaves in the trees and bushes, or under stones or dilapidated buildings.

The little white-footed or deer-mouse ranges over the whole United States. It is a pretty little creature, living in the fields, and making its nest of grass and leaves in the bushes. Several species of mice are known to 'sing,' producing a note much like that of a canary. By some this song is supposed to be due to some bronchial or other affection; but this is certainly not proven. Dr. Lockwood had a white-footed mouse which was a

great singer, and his account of its accomplishments is a delightful bit of
mouse biography. He has set the songs to music, and given them names.
He describes one performance in the following words: "She was gambolling in the large compartment of her cage, in a mood indicating intense
animal enjoyment, having woke from a long sleep, and partaken from
some favorite food. She burst into a fulness of song, very rich in its
variety. While running and jumping she rolled off what I have called
her Grand Role; then sitting, she went over it again, ringing out the
strangest diversity of changes by an almost whimsical transposition of the
bars; then without for an instant stopping the music, she leaped into
the wheel, started it revolving at its highest speed, and went through the
Wheel Song in exquisite style, giving several repetitions of it. After this
she returned to the large compartment, took up again the Grand Role, and
put into it some variations of execution that astonished me. One measure,
I remember, was so soft and silvery, that I said to a lady who was listen-
ing, that a canary able to execute that would be worth a hundred dollars. . . . So the music went on, as I listened, watch in hand, until
actually nine minutes had elapsed."

The elegant little dormice of the Old World seem to connect the
squirrels and the mice, being like the former in their habits, and more like
the latter in structure. They live in the trees, running and jumping with
as much grace and agility as our squirrels, or sitting up on their haunches,
eat their nuts in exactly the same way as do the others. But when winter
comes, the dormouse becomes torpid and so lethargic as to have given rise
to a proverbial expression, — 'sleeps like a dormouse.'

Before speaking of the true beaver, a few words must be said of the
giant fossil beaver which once inhabited the forests of the New World.
In structure it combined the characters of the beaver and the chinchilla;
in size it equalled a full-grown black bear. It was contemporaneous with
the mastodon; but no one knows anything about its habits. But few
specimens are known, less than a dozen being enumerated in the last
paper on the subject. It ranged from South Carolina and New York to
Michigan and Texas.

The beaver is by far the most important of all the rodents, and as such
demands more space than we have accorded to any other member of the
group. There is but a single species in the world, and this inhabits the
northern portions of both continents. Between the beaver of Europe and
that of America there are some slight differences; but not enough to make
them distinct species. At the time of the settlement of North America
the beaver supply of Europe was at a low stage, and hence the opening up
of a new source of skins was very important. At that time the beaver

extended from Florida north to the limits of trees, and from the Atlantic
west to Arizona, California, and Alaska. There were beaver everywhere.
It is no wonder then that the opening up of our shores and inland waters
gave a new impetus to trade; and, in fact, to the beaver is due a large
portion of the settlement of New York, Canada, and the west. One has
but to read the pages of Parkman to see how all-important this animal was

Fig. 485. — Beaver (*Castor fiber*).

to the colony of New France, and how several times the community was
in danger of complete wreck because the hostile Indians prevented the
beaver-skins from reaching Montreal and Quebec.

Some of the old superstitions connected with the beaver deserve mention. Near the vent are two glandular pouches, which secrete a strong-
smelling substance known as castoreum. In olden times the beaver was

largely hunted to obtain this substance, and the old hunters used to declare that the animal knew their object, and that when pursued, he would gnaw off the pouches, leaving them behind for the hunter, and thus save his life. This castoreum was highly esteemed as a remedial agent, and a list of the diseases it would cure is about equivalent to a complete catalogue of all the ills that human flesh is heir to. The beaver is pictured in all the older works as a marvel of knowledge, and the beaver communities as wonderful social organizations. Says Tonti, who accompanied La Salle down the Mississippi over two hundred years ago, every beaver " is obliged to work; but if any one has his tail excoriated or otherwise hurt, he lays it flat upon his back, to show that he is unable to work." Tonti gives a pretty tolerable picture of a beaver community; but the real facts will hardly bear him out when he says, " they make a canal or subterranean aqueduct from the river to one of the apartments, in which they have a kind of pond, wherein they hold the tail; otherwise they could not live."

Some of the other side, the reality of beaver life, now needs mention. To-day but a few isolated communities can be found in the United States east of the Mississippi, but in Canada and the northwest it has not yet been exterminated. Elsewhere every specimen is caught as soon as found, the white hunter paying no attention to the old Indian law of leaving some in every pond to perpetuate the colony. Soon the few stragglers left in northern Maine, in the Adirondacks, and in Virginia will be gone.

The beaver, aside from its skin, is best known as an hydraulic engineer and for the formation of its dams. In these respects it does not always act as some accounts would indicate, for sometimes its instincts are at fault; but in general it displays an astonishing amount of skill in its work. First comes the building of the dam, — a heavy, solid structure reaching across some small stream. These are built by felling small trees and underbrush with the sharp, strong, chisel-like teeth, and placing them in the desired position, thus making a foundation. To this are now added twigs, dead leaves, and mud, until a solid, water-tight structure is the result. This dam is to make a pond in which the houses are placed. These are rounded or oblong structures of brush, standing in the water and covered with mud, and with the entrance or door below the watermark. The principal use of the pond is to afford an easy means of getting their food to their houses. They eat the bark of various trees, and to obtain it they cut down trees, sometimes of great size. This is done entirely by gnawing round and round until at last the tree gives way; this process results in giving both the log and the stump a conical shape, as shown in the cut. After the tree is down, the branches are lopped off.

and the trunk cut into suitable lengths, and then all is hauled to the neighborhood of the houses to form the winter's supply.

In trapping beaver considerable skill is necessary. The castoreum is rubbed on the trap not only to attract the beaver, but to kill the human smell. The trap is baited with some favorite wood, like swamp-maple or mountain-ash. Then, too, the trap must be fastened, and were the trapper to tie it to any tree, the chances are that the trap and the beaver would both be missing at the next visit, while the stump would show exactly how the operation was done. The trapper knows, however, that the beaver will not gnaw a dry spruce stake, and hence he fastens the trap to one. When caught, the skin is taken off and cured. A beaver-skin will weigh from two to three pounds, and is worth at first hands from two to three dollars a pound. The flesh is good, and the meat of the scaly, paddle-like tail is esteemed as an especial delicacy.

In Oregon and Washington Territory occurs a strange rodent known by a name equally strange; it is the sewellel or showt'l; but besides, it has other cognomens like mountain-beaver, mountain-boomer, ground-hog, gopher, badger, and shote. The first two names are corruptions of the Indian words for this animal. It is emphatically a burrowing form, and in the broken country on the western base of the Cascade range the ground is riddled in places with its holes. These communicate with burrows of vast extent, and are as much of a nuisance as those of the gophers. It lives on various vegetable substances, and after gathering the plants they are said to leave them on the logs to wilt and dry. The Indians in their myths believe this to be the first animal created with life. To the naturalist it is interesting, as it combines the structure of the marmot and of the beaver with some peculiarities of its own; to the Indian it is of importance, as it furnishes both food and clothing, its skin being much like that of the muskrat.

The marmots embrace a considerable number of forms more or less like our common woodchuck, which is one of the largest members of the group. It is, however, too well known to demand farther notice, especially as it falls far short of its celebrated relative, the prairie-dog, in interest. All over the western plains to the Pacific slope occur their villages, sometimes of immense extent. Sometimes for miles these hillocks dot the prairie, each little mound of earth forming an observatory for the curious little animal who, sitting bolt upright upon it, watches every movement, and then, when his curiosity is satisfied, gives a comical and almost derisive flirt of his little tail, and disappears headlong down the burrow beside him. But the curiosity returns, and in a few moments a venturesome nose and a bright pair of eyes appear at the mouth of the hole, exhibiting a perfect picture of inquisitiveness tempered with a due amount of caution. The prairie-dogs are constantly digging. No matter how large their underground nests may be, they are constantly trying to make them larger, and so day by day the hillocks grow, while the subterranean passages become interminable.

These animals are frequently tamed and domesticated, and most amusing pets they make. They will make their burrows in the yard; and though so tame as to come regularly for their meals, they always show their native

FIG. 486. — Prairie-dogs (*Cynomys ludovicianus*).

traits as well in domestication as in their wild condition. They show the same desire for soft linings to their nests, the same timorousness except at meal-time, and they dig as well and as constantly as if on their western prairies. One has, however, to be careful; for these animals are great gnawers, and nothing except articles of stone or metal are safe from their teeth. We have already alluded (pages 389 and 527) to the alleged communism which is stated to exist between prairie-dogs, owls, and rattlesnakes, and need not refute it again. It is, however, a question whether these animals ever drink in their own homes. In confinement they do occasionally, but they have been known to live for six or eight months without water, except such as they might get from their food and the dew of the evening.

The ground-squirrels are a connecting link between the marmots and the true squirrels. The most marmot-like are some little forms — Spermophiles, the books call them — which have no distinctive common name, but which share the word gopher with the forms already enumerated. In general, a description of their habits is about equivalent to what we have previously given on the pouched gophers and the prairie-dogs, but we must make room for an extract of another aspect of these animals from the pages of Dr. Coues. " A gopher must be seen, and seen often, to be appreciated. For instance, a gopher caught away from home is a very different animal from one at the mouth of his hole. A most unreasonably timid animal, considering how rarely he is molested, he never goes out without feeling that he has taken his life in his hands. A thoroughly scared gopher is the liveliest object in nature ; a mule kicking over the traces is perfect repose in comparison. He doubles up and opens out like nothing else I know of, with his absurd little whisk of a tail hoisted, and the way he gets over ground without once looking back is amazing. Safe home, be he never so frightened, he will stop to see what was the matter. He pops bolt upright, stands stock still with his fore paws drooped affectedly in front of him, looks demurely around, and squeaks out, 'Pooh! who's afraid?' as plainly as possible. But let one come a step nearer, and down he goes on all fours, right over the hole, where he sits and scolds, with back arched up, ready for a dive. When he does finally duck out of sight, there is no mistaking his meaning: the suggestive flirt of his tail, the last thing seen, speaks volumes to a thoughtful observer. . . . But the prettiest of all the exhibitions a gopher can make of himself is when he frames his profile in the rim of his burrow. Not seldom, after running some little fellow to earth, have I stood still, just by the hole, and confidently waited for his reappearance. Presently I hear a little scratching, perhaps a squeak, and then I see his head, turned roguishly to one side, to throw one black eye full upon me, as if to ask what manner of creature I may be, to stand thus boldly at his door. He looks as if he would like to invite me in, and then laugh at me for being too big and too clumsy to enter."

In the pretty little ground-squirrels — chipmunks, every one calls them — we are a step nearer the true squirrels ; pretty and graceful they are as they sit perched upon the stone wall, or run along the rails of a fence, their reddish fur ornamented with stripes of white and black. Every boy knows them thoroughly. At the roots of yonder tree is a small hole, dug by this little creature; and if you dig down and follow it a little way you will find a warm nest of grass and leaves, and perhaps you may find his store of beechnuts and other provisions, laid up for a time of scarcity. There, as you dig, comes the chipmunk home. He sees you and stops. His cheeks are filled with seeds, which he was about to add to the stock you have rifled. You step towards him, and down he goes on the other side of the wall, to disappear in the grass.

The true squirrels are larger and more arboreal in their habits, and in these we find that full development of the tail which gives these animals their scientific name. *Sciurus* was the name the poetic Greeks gave to this animal; it means shade-tail, and our word squirrel is plainly a deriva-

FIG. 487. — Flying-squirrel (*Sciuropterus volucella*).

tive of it. Isn't the name apt? See the squirrel as it perches on the limb of a tree, holding a nut in its paws with almost human readiness and facility, with his bushy tail arched up over his back, the tip gracefully curved away from the body, and the full force of the name will be at once apparent. It now appears that there are six species of true squirrels in the United States, but in the older work of Audubon and Bachman twenty-four are enumerated. This will show how variable these animals are in their appearance. Of their habits there is little to be said that is not known to all.

Of all the squirrel tribe there is none so pretty or so interesting as the little flying-squirrel, or assapan. It was one of the earliest observed animals of our country. John Smith, in 1606, mentions it. and a little later Morton. in his 'New England Canaan,' speaks of "a little flying squirrill. with bat-like winges. which hee spreads when hee jumpes from tree to tree. and does no harm." It is a pity that these little creatures are not more abundant. for they truly do no harm; but, on the other hand, they are about as pleasant companions as one could wish in the trees about his house. During the day they are mostly quiet, and it is only as night comes on that they show themselves in all their grace and beauty. Occasionally they seem to have especial times of play; when it appears that they were so filled with animal spirits that they must seek every means to let it out. At such times a grove inhabited by these animals presents a most interesting sight. Each squirrel is in full motion, now climbing up some tall tree. and then, with legs spread out, stretching the 'wings' on either side of the body, he gives a spring out into the air, and sails away. At first he falls rapidly. but as he gains more momentum, he glides away farther and farther, until at last he alights. at a much lower level, but at a distance of even ten or a dozen rods from his starting-point, on the trunk or lower branches of another tree. Up this he quickly scrambles, to repeat the sail in some other direction. This flight seems to be indulged in as pure recreation, and at these times it seems to have no other object.

The flying-squirrel itself deserves a word of description; our illustration of it is most admirable. It is flatness itself; its body is flat, and then when it expands those folds of skin which extend between the legs of either side, this appearance is increased. And then there is the tail, flat like the rest, the hairs running off on either side, like the barbs on a feather. The pretty grayish fur of the back, and the white of the belly, is very soft, and the large, bright, black eyes add not a little to the beauty.

In captivity these animals make the most engaging pets. They soon become most thoroughly tame; and they never seem to show the temper and the teeth as the other squirrels do. One which I had liked to snuggle up inside my vest, making his entrances and exits by way of my coat-sleeve. Here he would stay by the hour, while I was walking about, and at times his head would peep out as if to see what was going on. He had his regular nest of cotton in a box. and here he spent most of the time during the day. He would. however, come out and climb to the shoulders of any one who chanced to be in the room, and take the same flying leaps as do his brothers in the woods. When coiled up in his nest he slept soundly. and it was only with difficulty that he could be wakened, and no matter how much one might poke him. he never showed the slightest temper. At night. however, he was fully awake, and then he showed all his inquisitiveness. It was a pleasure to watch him take the nuts from one receptacle, and then hunt for some other place to hide them. If they were placed in his nest, he seemed to think my pocket was a far preferable place to store them, while the pocket in turn was deemed far less secure than a hole underneath the sofa. In every motion he was a picture of grace, and every one was soon his friend.

Insectivorous Mammals.

The small animals known as insectivores need detain us but a little; for they are comparatively uninteresting. Indeed, a large number of them possess not the slightest interest for any except the professional naturalist. As their name implies, insects form the larger part of their food; but this is merely a physiological character, and is of no consequence in classification. With this peculiarity in food there are also a number of structural features which at once mark the group as distinct from all other mammals. It must be remarked, however, that all the members are not exclusively insectivorous; the tanrec of Madagascar feeds on earthworms, the almiqui of Cuba will attack poultry, another Malagassy species destroys rice, while others may eat fruit, fish, and various other substances. These are, however, but exceptions to the general rule.

The hedge-hogs and the porcupines are frequently confused, in the common mind, but in reality the two are very distinct, and have little besides the armor of spines in common. But even in these we notice a difference. In the porcupine the spines are easily detached, and may be left sticking in the mouth of the dog who is so venturesome and rash as to attack one of these animals. In the case of the hedge-hog, on the other hand, the spines are firmly implanted in the skin. When the hedge-hog is attacked, he merely rolls up in a ball, presenting nothing but spines sticking out in every direction. One species is common in the Old World, extending from England and Ireland to China, and the other species, eighteen in number, are likewise confined to the Old World. The species figured seeks some hole in which to spend the winter, either under the roots of trees or in clefts of the rocks, and here it stays in a completely torpid state until the return of spring.

The shrews are far more familiar to us than the spiny creature just mentioned. They are very mouse-like in appearance, except that the nose runs off into a long, sharp point. Some of our species are terrestrial, and some aquatic, but none have attained the notoriety of the common species of England. Around these an extensive folk-lore has accumulated. One belief is that if a shrew but cross a path or road, instant death is the penalty. Again, if a shrew touch another animal, the part touched is sure to be affected with disease unless the precaution were taken to employ some equally superstitious remedy. This is referred to by Gilbert White, in his 'Natural History of Selborne,' possibly the most celebrated book on natural history ever written. "Now," says White, "a shrew-ash is an ash whose twigs or branches when applied to the limbs of cattle will immediately relieve the pains which the beasts suffer, from the running of a shrew-mouse over the foot affected; for it is supposed that the shrew-mouse is of

Fig. 488.— European hedge-hog (*Erinaceus europeus*).

so baneful and deleterious a nature that whenever it moves over a beast, be it horse, cow, or sheep, the suffering animal is afflicted with cruel anguish, and threatened with the loss of the use of the limb. Against this accident, to which they were continually liable, our provident forefathers always kept a shrew-ash at hand, which, when once medicated, would maintain the virtue forever. A shrew-ash was made thus: into the body of the tree a deep hole was bored with an auger, and a poor devoted shrew-mouse was thrust in alive, and plugged in, no doubt with several quaint incantations, long since forgotten." One of the shrews of southern Europe is to be noticed; for it is barely two inches in length, and is the smallest mammal known.

The moles are the best burrowers of all the insectivores, and not one of all the mammals exceeds them in adaptation for a life under ground. Their fur is thick and soft, and the small bead-like eyes are concealed by the hair so that the dirt cannot get into them; and then the broad, long-clawed fore feet are the most efficient digging-organs. Almost all of their life is spent under ground, and their galleries run in every direction. The habits of our native species have not been carefully studied, but it would appear that none of them are equal to the European species in their abili-

Fig. 480. — Common mole of Europe (*Talpa europea*).

ties as sappers and miners. The common European species figured is the best known and most celebrated of all. Many naturalists have studied its habits, and there is such a uniformity in their results that we must believe their accounts, wonderful as they may seem.

The central portion of the ranges of the mole is in a small, hard hillock, raised by the moles, and rendered firm and strong by the earth being packed solidly together by the animals when forming it. Inside this hill, at the base, is a circular gallery, and from it five passages go obliquely inwards and upwards to another circular gallery near the top of the hill, and from this three other passages go down to the chamber in the centre of the mound — the home of the mole. From the first-mentioned circular gallery the passages radiate in all directions, but each

eventually comes back to one principal road which leads from the central
hill, or 'fortress,' to the most distant part of the mole's range. From this
principal road the mole runs out drifts in every direction in his daily
search for food. This main road is made by compression of the earth, but
the lateral branches are constructed rapidly by throwing up the earth at
intervals to the surface, — a true excavation. The mole, in running these
branches, finds worms, insect larvæ, and the like, on which it feeds. The
mole, in order to go from his dwelling to his feeding-grounds has to trav-
erse this main road, and eventually becomes very smooth. In passing
through it he usually makes pretty good time, and the experiments of
Le Court at the beginning of this century, as related by the late Thomas
Bell in his History of British Quadrupeds is worth quotation.

"Having ascertained the exact direction of the road, and finding that
the mole was engaged in exploring for its food the ground at the farthest
extremity from the fortress, he placed along its course, at certain distances,
several pieces of straw, one extremity of which penetrated within the
passage, and to the other end was fixed a little flag of paper. He also
introduced into the passage near the end a horn, with the mouth-piece
standing out of the ground. Then, waiting till he was sure of the mole's
presence at that part of the road, he blew into the horn, to use the words
of Geoffroy, 'un cri effroyable'; when, in a moment, the little flags were
successively thrown off, as the mole, in its rapid course toward its fortress,
came in contact with the interior extremities of the straws, and the spec-
tators of this neat and demonstrative experiment affirm that the speed of
the frightened mole was equal to that of a horse at full trot."

Of our American moles only one need be mentioned; this is the curious
star-nosed mole, which receives this name from the star-like appearance of
the tip of its nose. Surrounding the nostrils are a number of delicate
fleshy filaments, each richly supplied with nerves, the whole giving much
the appearance of a star: whence the name. In the nose of the common
mole the nerves are also well developed, but in this form we find a much
further specialization. It is easy to see what purpose these fulfil in these
animals. They are of great use in ascertaining the presence of suitable
morsels of food in the soil, and in places where the eyes could not be of
the slightest use.

It is a considerable step from the habits of the moles to those of the
animal shown in the next cut, the colugo, or kaguan, of the East Indies,
a form much more like a flying-squirrel or a bat in general appearance
than like a mole or hedge-hog. The animal itself is not well known.
All the points in its structure show that it is a true insectivore modified for
an arboreal and aerial life in much the same way as is the flying-squirrel.

It has a broad membrane extending from its neck all round the body, and reaching to the extremities of the toes and of the tail, so that when the limbs are spread, the whole surface presented is very large, and hence the animal is able to glide even better than the form with which it has been compared. For our knowledge of its habits we are mainly indebted to Wallace, who observed them in Sumatra. The colugo, or kaguan, as this animal is called, rests during the day, clinging to the trunks of trees,

Fig. 490. — Kaguan, or colugo (*Galeopithecus volans*).

where its mottled fur resembles closely the bark, and no doubt aids in protecting it from observation by its enemies. In the twilight it is more active, and it leaps and sails from tree to tree like a flying-squirrel, sometimes passing a distance of seventy yards. Wallace states that the animal lives on leaves, — a strange diet for an insectivore, — and the structure of the stomach and intestines would tend to confirm this, though the strange character of the teeth would indicate that there was some other element in the diet; perhaps insects are also largely eaten.

BATS.

The bats are the only true flying mammals. The colugo, the flying-squirrel, and the like do not really fly; they merely slide down an inclined plane of air, and their 'flight' always terminates at a lower level than that which they started from. It is true that they can rise above the lowest part of the course; but this is due not to any volant powers, but to the momentum they gained in their previous descent. In the bats, on the other hand, we have flying powers as well developed as in the insects and birds. Here true wings are found which need a moment's attention.

Such a thing as a bat's wing is not paralleled in the whole living animal world. Its thin, skinny appearance is very disagreeable, and it is notable that when artists wish to picture a griffin or goblin, or other horrible creature of the imagination, the wings of a bat are almost invariably appended to complete the idea. A glance at our cut will show some features of interest in relation to these wings. In the birds the wing is formed by stiff feathers borne on the arm, and the bones of the hand are very degenerate. In the bat, on the other hand, the wing is but a membrane which needs some more solid support, and for this purpose the bones of the hand are highly developed. The fingers especially are very long, and four of them form long rods, having the same relation to the membrane of the wing as the ribs of the umbrella have to the cloth. One of the fingers, however, does not enter into the support of the wing; but rather serves as a hook or claw — the only substitute for a hand.

Bats are nocturnal animals, and in our summer evenings, soon after dusk, they appear, sometimes in large numbers, flitting about on noiseless pinions in their search for mosquitoes and other insects of evening. They are apt to inspire feelings of dread; but needlessly, for there is no danger: they will not touch any one or do the slightest damage. Even in the darkest places they will steer clear of every obstacle, not touching the smallest thing. It is not to sight alone that they owe this skill; the wings are very sensitive, and are richly supplied with nerves, so that they can feel even the slightest changes in the air — changes which we, with our much duller senses, can hardly conceive of. Most of the bats are to be regarded as beneficial in that they feed on insects, and devour immense numbers of those unmitigated nuisances, the mosquitoes. None that we have with us do the slightest harm, though if caught in the hands, their sharp, needle-like teeth will draw blood from the fingers.

All of the bats, however, are not insect-eating. There is a group of large forms living in the tropical parts of the Old World which feed solely on fruits. One of these forms is shown in the adjacent plate. From

FLYING-FOXES (*Pteropus edwardsi*).

its large size, and the shape of its head and body, it is known as the flying-fox. All the species are much alike in their habits. They rarely exceed five feet in spread of wing. In the morning some of them may be seen between the hours of nine and eleven; then they retire to the woods to spend the heat of the day, and again at evening they come out in much larger numbers, and frequently do considerable damage to the fruit-groves and plantations: mangoes, guava, plantains, and other fruits suffer, and, says Thwaites, they frequent the places where palm-toddy is being brewed, and get just as disreputably drunk as any human being on the same liquor. In the early hours of the morning they may be seen flying home on unsteady pinions, in a state of riotous intoxication. They, however, have no occasion to let themselves in with a latch-key; for their habitation is of a very primitive character, and is unprovided with locks and bars. All that they need is a forest where they will be undisturbed, and here, hanging head downwards, their hind feet grasping the limb, they sleep off the effects of their debauch. When thus hanging, they wrap their wings about them, and whenever I have seen them hanging thus in captivity, I have invariably thought of the stage villain wrapped for concealment in his cloak of black.

These bats have an economic importance in the lands where they are abundant, for they are extensively used for food. The natives go in the day-time to their roosting-places, knock them down with sticks, and carry them home by the basketful. In preparing them for the kettle or the oven, great care is necessary, as the fur and skin has a very powerful 'foxy' odor, and the least contact of this with the flesh is sure to communicate the flavor. They are usually cooked with an abundance of spices, and are said to taste much like hare. One or two facts more must be mentioned. In their flight they resemble crows rather than bats; and Haeckel, who saw them in Ceylon, refers to their drinking habits and the cause, saying that "this predilection may no doubt be amply accounted for by the near affinity of the bats to the apes,—as proved by their phylogenetic pedigree,—and through apes to man." Haeckel is nothing if not an evolutionist, but this is a species of evolution not laid down in the works of Darwin, Lamarck, or even Buffon.

The rest of the bats live upon animal food, and insects especially form the diet of most of the group. They are even more nocturnal than the fruit-bats. At night they fly about, but in the day-time they retire to some dark place to sleep away the hours of light. Sometimes this is in some cavern or hollow tree, but frequently some deserted house is taken for a home, and instances are on record where thousands have made their headquarters in the attic of a dwelling, emerging after night to the great annoyance of the inhabitants, and requiring no little labor to clear the premises of them. Observations on their habits are few; were these animals better known, we should doubtless find that they presented many and interesting differences; but there is a great similarity in the broader features, so that a general description of one will answer for all.

A little brown bat which I kept some time in confinement was at first very pugnacious when first taken, squeaking and trying to use its small but sharp teeth when tormented, but after a few days it became more reconciled. I fed it at first with minced beef and an occasional fly; but one day I found that a bit of beef left in the cage had begun to decay, and this attracted flies, acting in fact as a trap, of which the bat took full advantage. As a fly entered the cage, the bat would knock it down with its wings, and then with these members spread would settle down over it so as to prevent its escape. Then the head would be inserted beneath the membrane, and in a few seconds would be brought out with the fly in the jaws.

Bats carry their young about with them. In some this is accomplished by the baby bats clinging to the fur of the mother, while in another, which has a naked body, nursing-pouches for the young are developed. When resting, the mother folds her wings about the young. When the young, at least in some species, are born, the tail is drawn upwards, making a bag of the membrane, which unites this member with the hind legs, and in this the little ones are caught and kept from dropping to the ground. Mention should be made of the peculiar appearance which bat hairs present under the microscope, each species having its own distinctive features. These hairs are well known to microscopists, and form a part of the collection of every amateur.

The blood-sucking habit of some bats has been a subject of much discussion. For a time it was thoroughly believed, and one species was selected as the embodiment of the tales. It was the species represented in our cut, and it received in consequence the name *Vampyrus spectrum*, the ghost-vampire. This was a rank libel on this innocent fruit-eating form, and gradually the whole tale fell into disrepute. Then Darwin, in his voyage in the ' Beagle,' observed similar habits in a bat belonging to another genus, found in Chili. He says: " The vampire bat is often the cause of much trouble by biting the horses on their withers. The injury is generally not so much owing to the loss of blood as the inflammation which the pressure of the saddle afterwards produces. The whole circumstance has lately been doubted in England. I was therefore fortunate in being present when one (*Desmodus dorbignyi*) was actually caught on the horse's back. We were bivouacking late one evening near Coquimbo, in

Chili, when my servant, noticing that one of the horses was very restive, went to see what was the matter, and fancying he could detect something, suddenly put his hand upon his withers, and secured the vampire."

Mr. Bates also confirms the fact by observations which he made at Caripé, near the mouth of the Amazon. "The first few nights I was much troubled by bats. The room where I slept had not been used for many months, and the roof was open to the tiles and rafters. The first night I slept soundly, and did not perceive anything unusual, but on the next I

Fig. 494. — False vampire (*Vampyrus spectrum*).

was aroused about midnight by the rushing noise made by vast hosts of bats sweeping about the room. The air was alive with them; they had put out the lamp, and when I relighted it, the place appeared blackened with the impish multitudes that were whirling round and round. After I had laid about well with a stick for a few minutes they disappeared among the tiles; but when all was still again, they returned, and once more extinguished the light. I took no further notice of them, and went to sleep. The next night several got into my hammock; I seized them as they were crawling over me, and dashed them against the wall. The next morning I found a wound, evidently caused by a bat, on my hip. This was rather unpleasant, so I set to work with the negroes, and tried to exterminate them. I shot a great many as they hung from the rafters; and the negroes, having mounted with ladders to the roof outside, routed out from beneath the eaves many hundreds of them, including young broods. . . . I was never attacked by bats except on this occasion. The fact of their sucking the blood of persons sleeping, from wounds which they make in the toes, is now well established; but it is only a few persons who are subject to this blood-letting. . . . I am inclined to think that many different kinds of bats have this propensity."

Elephants.

There is scarcely another animal — man excepted — about which so much interest centres as about the elephants. Their immense size and their great intelligence render them favorites with all. And then they have so many interesting features, — their long legs, which seem to move without a joint; their small eyes, which seem all out of proportion to the huge body; the ivory tusks; and, strangest of all, their trunk, which words can scarcely describe.

There are two distinct species of elephants, — one living in Asia, and characterized by small ears, a hollow forehead, four nails to the hind feet, and tusks in the male only; the other is the African elephant, with a convex forehead, enormous ears, and only three nails on the toes of the hind feet. In the structure of the teeth as well there are differences to note. In all elephants there are but two kinds of teeth, — molars and incisors. The latter are represented in the upper jaw alone, and these form the tusks, so familiar to all. These grow from permanent pulps, and hence they increase in size as long as the animals live. The molars are complex in structure. They are enormous in size, and each one is made up of plates of enamel, placed perpendicularly to the surface in use, while between and around these plates is the softer dentine. As a result of this the cutting surface of the teeth, by wear, soon become converted into a series of ridges (the enamel) and depressions, and thus the two opposing teeth of either side form a most perfect mill for the comminution even of the branches of the trees. The enamel ridges in the Indian elephant are straight and narrow, while in the African they are fewer and lozenge-shape.

There is a curious succession of these molars in the elephants. There are twenty-four in all; but of these only eight — two in each side of each jaw — are in function at once. As the tooth comes down from the jaw, it comes into use, and then slowly moves forward, pushing its predecessor along. and at last forcing it out of the mouth, and, in turn, it is itself forced out as soon as it becomes worn out. In this way there are **three** gradual changes of the teeth.

The trunk is a wonder. One can stand and watch these animals for hours. and continually find new points to excite his amazement in the marvellous capacities of this Brobdignagian nose; for it is in reality but the nose of the elephant, highly specialized and greatly prolonged. How wonderfully it plays the part of a hand! The little finger at the end is far more flexible than any thumb, and it is endowed with a very delicate sense of touch. With the trunk the animal feeds itself, pulling up the grass. or tearing down the branches of the trees, or again, perhaps. taking the peanuts and cakes of the circus. As it sticks out in the most beseeching way it seems to have a life of its own, and to be something entirely distinct from the huge creature behind. It seems to find its way to the food by instinct; for no one can realize that the small eye, way back in the distance, has any part in ascertaining what is within reach.

Its use in drinking is also wonderful; the way in which it becomes filled with water, which is then poured down the throat, is always a strange sight. no matter how often it has been witnessed. And then when the elephants take their bath, what a wonderful shower-bath the trunk makes! The water, which at other times is poured into the mouth, now falls upon the back. or perhaps is directed against some on-looker who has aroused the ire of the monstrous beast.

The size of the elephant has always been a subject of dispute. How large they may grow no one knows; but the adult Indian elephant has an average of about ten feet, the African being a little taller. Tales are told of elephants much taller than this, but the following quotation from Sir Emerson Tennent's 'Ceylon' explains some of the accounts: "Elephants were measured formerly, and even now, by natives, as to their height, by throwing a rope over them. the ends brought to the ground on each side, and half the length taken as the true height. Hence the origin of elephants fifteen and sixteen feet high. A rod held at right angles to the measuring-rod. and parallel to the ground, will rarely give more than ten feet. the majority being under nine."

The most interesting aspect of the elephants is that exhibited by their intellectual side. According to the cut-and-dried systems of anatomists the intelligence of the elephant should fall below that of many other animals. while in reality it stands far above all but a very few. The pages of natural histories and the records of travel are filled with illustrations and anecdotes. showing their wonderful intelligence and sagacity, some of which. fully authenticated, would seem to indicate a degree of intelligence

AFRICAN ELEPHANTS (*Elephas africanus*).

equal to that exhibited by some individuals of our own species. Still, we must constantly bear in mind that tendency to exaggeration which forms a part of every traveler's luggage. Memory and sympathy are well developed, while revenge is only paralleled or exceeded in the monkeys. They have, too, an understanding, and readily learn to perform complex acts, but nowhere does the intellect of these animals show itself more than in the case of the decoys used in capturing the wild elephants. For the following account we are indebted to the work of Tennent just quoted from, his details, however, being somewhat condensed.

In Ceylon wild elephants are captured by driving them into a corral formed of trees, interlaced with huge timbers, the whole forming a fence so strong that not even an elephant can break through it or tear it down. The wild elephants being safe inside, two decoys were driven in, each bearing its driver, and the most intelligent one the chief nooser, while the other, a very strong animal, performed the heavier work. The first elephant moved along slowly, with an assumed air of indifference. As she approached the herd, the leader advanced towards her, passed his trunk gently over her head, and then went back to his companions, the decoy following and standing close behind him, so that the nooser could slip the rope around the foot. The danger was perceived, the rope was shaken off, and the wild beast would have punished the man had not the decoy interfered and driven him into the herd. The two decoys now pushed in and stood one on either side of the largest animal, while the nooser placed the rope, one end of which was fastened to the collar of the decoy, around the leg. The two tame elephants instantly fell back, one dragging the captive, while the other prevented any interference from the rest of the herd. In order to tie him to a tree he had to be dragged backwards some thirty yards, bellowing in terror, plunging on all sides, and crushing the smaller timber, which bent like reeds beneath his clumsy struggles. At last the rope was wound around the proper tree, but with a coil around the trunk the decoy could not haul the prisoner close up; so the second tame one, perceiving the danger, left the herd, and pushing the captive head to head, forced him to the tree, while the other hauled in the slack; and then the noosers hobbled the other legs.

With the others of the herd, similar tactics were employed: when an elephant seized the ropes with her trunk, to sever them with her teeth, a tame elephant pressed them down with her foot. The tame elephants displayed the most perfect conception of every end to be obtained, and the means necessary to accomplish it, and enjoyed all that was going on. There was no malignity of spirit in the heartless proceeding. Their caution was as remarkable as their sagacity; they never were confused, never hurried, never tripped on the ropes or got in the way of those already confined. One could almost fancy there was a dry humor in the way the decoys played with the fears of the wild herd, and made light of their efforts at resistance; they drove them forward, forced them back, forced them to rise when they lay down, or when it was necessary to keep them down, knelt upon them to keep them from rising until the ropes were secured.

There is another side to the elephant besides that detailed in most works. Usually the elephant is pictured as a mild and gentle animal; but in reality, even among those in our menageries, there are many instances where the whole apparent docility is the result of the fear of the keeper's prod. Jumbo had one side for the public, and another for private exhibitions. Elephants are ruled not by love, but by fear; a mouse will frighten them and even put them in a perfect terror. The keeper must constantly keep them in subjection; and if they see the slightest exhibition of fear on his part, the papers soon chronicle the fact that the ' Emperor' accidently killed his master. The picture of this side of the elephant's character which is drawn in Charles Reade's ' Jack of All Trades ' is not greatly overdone.

In most cases the fossil members of a group can have but little popular interest, but the extinct relatives of the elephant are too often mentioned to be omitted here. Probably no extinct animals are more familiar to all, at least by name, than the mammoth and the mastodon, and there is every reason to believe that both have become extinct since the advent of man upon this planet. The mammoths were most like the elephant in structure, having the same general character of teeth; while the mastodons had those instruments of mastication built on a somewhat different plan, and their grinding surface presents a strange arrangement of comparatively large hills and valleys. Of the several species of mammoths, none has acquired more celebrity than the Siberian species. It was a hairy beast, living no one knows exactly how recently in that land of snow. At the close of the last century a hunter found the carcasses of some of these animals imbedded in the ice, the flesh being in so fresh a condition that the dogs would eat it. With us mastodon remains are far more abundant than those of mammoths, and of the latter but rarely is anything except the teeth found. Mastodons, however, have become so common, that, except in the case of unusually perfect specimens, only newspaper record is made of the fact. The remains are usually found in the peat of some swamp, and the inference is that the animal became mired, and died where the bones are found.

THE CONIES.

In several places in the Old Testament the word 'shapan' occurs, a word rendered in our English version by 'cony.' This little animal is not described with scientific precision, and hence it has made no little trouble for commentators. For a long time it was supposed that it referred to some of the hares; but to-day all are agreed that it is the little animal figured here. From the various references found in Leviticus, Deuteronomy,

FIG. 492. — Cony (Hyrax syriacus).

Proverbs, and Psalms, we find that the conies were a 'feeble folk,' making their homes in the rocks, that they were small but wise, that their hoof was not divided, and that they chewed the cud. All of these characters and characteristics except the last are perfectly applicable to the form now identified as the cony. This animal, however, does not really chew the cud; but when at rest, it keeps its jaws in such constant motion as to convey the impression that it is chewing the dinner clipped some time before.

Except for this biblical reference the cony would possess but little of popular interest. It is a small and timid animal, making its home among the rocks, and feeding upon vegetation, but especially preferring the young and tender shoots of trees and shrubs. When the colony is feeding, a sentinel is posted to give warning of any approaching danger, and at the first alarm the whole herd quickly make their way to the rocky fastnesses, from whence it is not an easy task to obtain them. In size the cony resembles a rabbit, but in little else. It is covered with a dark brown hairy fur, and its front teeth grow from persistent pulps. The various features of its internal structure have caused the naturalist no little trouble. At various times the cony has been assigned a position along with the rabbits and squirrels, and with the elephants; but now it is placed in a group by itself, its only associates being three African species belonging to the same genus (*Hyrax*), but which are not so well known as the Syrian form figured. One of these differs in habits from the rest, in that it makes its home in the hollow trunks of the large trees of western Africa.

UNGULATES.

The group of ungulates (the name means hoofed animals) is the most important group of the animal kingdom, viewed from the economic standpoint; for it contains the great majority of our domesticated animals, and furnishes the principal part of our meat supply, as well as a considerable proportion of our clothing and foot-wear. The horse, cow, pig, sheep, and camel are familiar members of the group. To the naturalist, as well, they possess an interest which can scarcely be paralleled in the whole animal kingdom; for in this group, thanks to the wonderful deposits of fossils in our western territories, the lines of descent and the methods of origin of the existing forms have been traced with an accuracy and a detail which is equalled nowhere else, and which cannot fail to convince any fair-minded person who looks at the evidence of the truth of the theory of evolution, a theory which to-day is accepted by every naturalist in the world. Indeed, so important are these animals in this respect that we cannot pass them by without an allusion to one point of structure — that exhibited by the foot. First as to the hoofs. These are of the same nature as the nails which tip our fingers, but modified to support the weight of the body, and to protect the tips of the toes from wear; for all these animals walk, not upon the soles of the feet, but upon the extremities of the digits. The primary number of these toes in the whole group of mammals is five, but in this group we find every number from one to five, and all can be traced back to the five-toed ancestor. No living form has that number of toes, but in the past such forms lived, and from them two lines have diverged. In the one the number of toes are even (two or four), while in the other they number one or three. This difference corresponds to two great divisions of the ungulates, and has been brought about by the degeneration of one or more of the toes. For instance, in the horse there is but one toe remaining in the functional foot (the middle one), but farther up we find on the side of the 'pastern bone' two small bones known as the 'splint bones,' which are respectively the second and fourth toes of the five-toed ancestor in a very rudimentary condition; and in rare abnormal cases a reversion to an older type may be seen, and one or both of these splint bones may bear hoofs. In the series with the toes two or four in number, the first toe is lost first, as in the case of the pig, and then the second and fifth go, as is seen in the great majority of the group. The various stages in this process are extremely interesting, but cannot be detailed here. We can only say that the odd-toed forms are regarded as lower than those in which the toes are even in number, and hence are to be considered first.

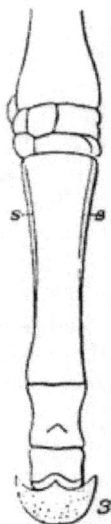

Fig. 493.—Skeleton of the foot of a horse. The middle toe (3) of the typical vertebrate foot alone reaches the ground; two other toes (the second and fifth) remain as the splint bones (S); the others have disappeared.

In the tapirs we meet a peculiar foot structure, which at once indicates their position at the base of the series. They have four toes on the fore feet and three on the hinder ones; a fact which taken by itself would render it difficult to say in which of the two groups of ungulates these forms should be placed. Fortunately there are other characters which decide the question. All of the tapirs with one exception inhabit Central and South America; the exception is found in the East Indies. They are rather stupid animals with nocturnal habits, living in the dense forests and eating all sorts of tender and succulent vegetable matter. Their flesh is good for food, and their hide makes good leather, and for these reasons they are hunted extensively. The most remarkable feature about them is the long nose (much longer in some than in the species figured), which is capable of considerable extension, and which can be used in much the same way as the trunk of the elephant.

There are six species of rhinoceros to be found in the tropical parts of the Old World. They are large and clumsy thick-skinned animals characterized among other things by the possession of one or two horns upon the snout. The one-horned forms are found only in the East Indies, while those with two of these nasal appendages are found in both Asia and

Africa. In the former the horn may grow to be three feet in length, forming a very efficient instrument as well as a weapon of offence and defence. In the two-horned forms the forward horn is usually much the larger, and may be four feet long, while the second may extend to half that distance. These figures are extremes: usually the dimensions are much less.

Fig. 494. — Tapir (*Tapirus americanus*).

The African species are much better known than those of India, and almost every volume of travel in the 'dark continent' details the excitement of a rhinoceros hunt. These huge beasts live in the dense jungle, lying still in the shade during the heat of the day, and going to their feeding-places in the morning and evening. At about eight at night the animal goes to the river to drink, usually following a regular path which he has broken for himself through the underbrush. A knowledge of this habit is of much use to the hunter, and it is in these paths that he awaits his game, or places his snares and pitfalls. There is far less excitement in hunting these animals than is the case with lions, elephants, or tigers, but

still a rhinoceros is an antagonist not to be despised. When brought to
bay he will charge, trying to impale horse or man upon the long, sharp
horns. The natives eat the flesh, while from the hide they make the
shields they use in their wars, and the horns are manufactured into sword-
hilts. Hunting them for the mere sake of killing them does not seem so

FIG. 495. — Rhinoceros (*Atelodus bicornis*).

praiseworthy an act as some travelers appear to think, and their details
of the struggles, charges, and final death, with perchance the slaughter of
a horse or two, possess but a morbid interest.

In the horses and their allies the reduction of toes reaches its extreme;
for in these forms the whole weight is borne on the tips of the middle toes
of each foot. Of course the most important of all is the domestic horse,

whose origin was probably in Asia, though in our western fossil beds bones are found which can scarcely be distinguished from those of man's faithful servant. The American stock, however, died out long before the discovery of this continent, and when the Spaniards first conquered the New World, no horses were known here. To go into the history of the horse and its various breeds would require a volume of no mean dimen-

FIG. 496. — Quagga (*Equus quagga*).

sions, while many pages could be profitably devoted to the mental aspects of the animal. We must, however, say a few words about the origin of the domestic horse from some wild breed — a point on which there is great uncertainty.

In the steppes of central Asia four varieties, or species, of wild horse are known, and one of these — a small dun-colored species with short, stiff, dark-colored mane — is regarded by some as the ancestral form — a view which receives considerable probability from the structural correspondences. There is, however, another side. Horses which escape from domestication and run wild change in their appearance, and we have reason to believe that in many respects they revert to their primal condition. Marked features of these are that they have a dark stripe extending along the back, two or three stripes across the shoulders, and transverse bars on the legs, — facts that would indicate that our horse descended from some form more like the zebra or quagga than the wild horse mentioned above.

Passing by the asses (descendants from the wild species south and east of the Mediterranean) and their hybrid with the common horse, — the mule, — we come to the striped horses of Africa, known as zebras and quaggas. Of these there are three species, all beautiful animals with their striped coats. The quagga is the least beautiful of the three, the stripes not extending to the hind quarters or the legs. They live in immense herds, and have a barking neigh, almost exactly reproduced in their name. The zebra is striped all over, and is one of the most beautiful animals in nature, but it is too familiar to need extended description or illustration ; specimens occur in every menagerie. The quaggas live in the low lands and plains, while the zebra inhabits the mountainous regions. They always post a sentinel whose sharp neigh sounds a warning when danger approaches, and at such times the whole flock after a moment's gaze dart off from their pastures into regions where they are safe from pursuit. It is frequently said that zebras and quaggas are untamable, but this is not correct. Still, it is a labor of some difficulty to break them to harness, and even when this is done the team lacks the reliability and steadiness of the old family horse.

In the even-toed ungulates we may have either four or two toes, and from what has already been seen those with the larger number must be regarded as the more primitive type ; but whether there be two or four, there is one feature to be noticed, — the foot is divided into two symmetrical halves, or, to use the Levitical expression, they cleave the hoof. The same law introduces us to another feature in these animals which may also be noticed here, especially since it serves to divide these forms into two minor groups. In the sumptuary laws just referred to it was not sufficient that the hoof be cloven, in order to render the flesh of the animal ' clean '; there was further necessary a chewing of the cud. In the pigs and hippopotamuses this does not take place, but in all the rest there is a rumination.

In these latter the stomach, instead of being a simple glandular sac, is divided into four portions, each with its special name and function. A cow, for instance, eats for a while, scarcely chewing the grass as it is taken into the mouth, but swallowing it together with a large amount of saliva. When first swallowed it goes to the first stomach, or paunch, and thence passes to the second division. In these it becomes wilted and softened, and then, when the animal is at rest, it is forced up into the mouth, to be chewed again. It now descends to the third and fourth stomachs, where it undergoes true digestion.

The true pigs all belong to the Old World. The domesticated animal appears to have originated from two distinct species. The pig of Europe and of our barnyards had its origin in the wild boar of Europe, while the domesticated swine of the Orient came from a species the home of which is in India. In later years these two distinct forms have been mixed, to the great improvement of our breeds. Only one feature needs mention in connection with our domesticated swine. The Levitical prohibition of pork was a sanitary measure, the full bearings of which have only recently been appreciated. There can be but little doubt that in the times that these laws were made trichinosis existed as it does to-day, and the observance of the fact that suffering and even death occasionally resulted from an indulgence in the flesh of swine led to the prohibition of the pig as a source of food supply. We now know that thorough cooking does away with all danger on this score. The wild boars of Europe and Asia (there are several species) are celebrated as furnishing much sport for the hunter. They are largely solitary in their habits, dwelling in the dense forests, and rooting through the ground in search of food of every sort. They are savage and courageous beasts, and their chase is attended with more excitement than one would imagine by looking into a pen filled with their fattened domesticated descendants. The wild species are thin, long-legged creatures, which run swiftly, and the long tushes of the males render them formidable antagonists for horses and dogs.

Of the strange 'pig-deer' of the Malay Archipelago, Wallace writes: "This extraordinary creature resembles a pig in its general appearance, but it does not dig with its snout, as it feeds on fallen fruits. The tusks of the lower jaws are very long and sharp; but the upper ones, instead of growing downwards in the usual way, are completely reversed, growing up out of long sockets through the skin on each side of the snout, curving backwards to near the eyes, and in old animals often reaching eight or ten inches in length. It is difficult to understand what can be the use of these extraordinary horn-like teeth. Some of the old writers supposed that they served as hooks by which the creature could rest its head on a branch. But the way in which they usually diverge, just over and in front of the eye, has suggested the more probable idea that they serve to guard these organs from thorns and spines while hunting for fallen fruits among the tangled thickets of rattan and other spiny plants. Even this is not satisfactory; for the female, who must seek her food in the same way, does not possess them. I should be inclined to believe, rather, that these tusks were once useful, and were then worn down as fast as they grew; but that changed conditions of life have rendered them unnecessary, and they now develop into a monstrous form, just as the incisors of a beaver or a rabbit will go on growing if the opposite teeth do not wear them away."

The only American pigs are the little, but by no means insignificant, peccaries of Central and South America, one of the species ranging north

FIG. 487. — Peccaries (Dicotyles torquatus).

as far as Arkansas. Unlike the wild boars, they form large herds which roam through the forests in their search for food. They are most courageous little beasts, or possibly it is better to say that they do not know enough to be afraid. When one of the herd is injured, the rest do not run away, but seek to kill the offender; and many a tale is told of human beings falling victims to the rage of these little animals, about three feet long and a foot and a quarter in height. Everything is afraid of them: even the puma and the jaguar, those lords of the forests, make way for a drove of these implacable little beasts. Their flesh is utterly useless as food, as it is thoroughly impregnated with the secretion of a gland opening on the rump, and which is described as being nearly or quite as offensive as that of the skunk.

Africa contains two species of hippopotamus, but the form whose countenance stares upon us from the cut is the best-known form. It is a huge form — fourteen feet long — which is utterly lacking in every element of beauty and grace. This clumsiness and ungainliness is not of so much account; for the animal spends most of its time in the water, which tends

FIG. 498. — Hippopotamus (*Hippopotamus amphibius*).

to buoy it up, so that an obesity which would be very inconvenient on land offers no special obstruction to the actual mode of life. It usually swims or floats with but little more than the nostrils above the surface, while it is capable of remaining below for a long time. It feeds upon vegetation of all sorts, and its enormous canine teeth are well adapted to rooting up the plants growing on the bottoms of the rivers and ponds. The hippopotamus is not an agreeable beast; indeed, it is one of the most dangerous animals in Africa. It is easily excited, and even the sight of a man will rouse it to an implacable frenzy. The natives, unless properly armed, let them severely alone; but when they do have a hunt, it is a celebration for all the neighborhood. The animal is taken in various ways in different regions, and the flesh affords the natives a good meal, while the fat is fried out into a lard which will keep for a very long time, even in the hottest weather, without becoming rancid. The flesh of the hippopotamus is not relished by most white people; some travelers praise it as a great delicacy, but others, apparently with more truthfulness, say that it is filled with sinews, and is scarcely more palatable or digestible than so much rope. The roasted gelatinous skin — two inches thick — is considered the greatest delicacy.

All the remaining Ungulates are ruminating, or cud-chewing forms, the peculiarities of which were referred to a short distance above. Of these the strangest members are the camels and their near relatives, the llamas and alpacas of South America.

The true camels are natives of the warm portions of the Old World, and have long been domesticated by man. Indeed, except in cases where they have escaped from domestication, and have returned to a condition of nature, no wild camels are known. There are two distinct species, — the common single-humped camel, or dromedary, and the two-humped, or Bactrian camel, the most characteristic features of which are the number of humps upon the back. These humps are really wonderful structures. One would suppose that they were supported by bone, but such is not the case; they are simply mountains of fat, and serve as stores of food for a time when food is scarce. In the regions inhabited by these animals there are two seasons, — the wet and the dry. In the former the vegetation is abundant and luxuriant, and then the camel has an abundance; but when the dry season comes, everything becomes parched, and the plants that survive are about as succulent, and scarcely more nutritious, than the paper on which this volume is printed. Then it is that the store of fat piled upon the back in the rainy season is drawn upon, and serves to tide over the time of scarcity.

The camel is a stupid and a vicious beast, about which no little sheer nonsense has been written. It is a continual grumbler, and at times is really dangerous to approach. Its face is one of the ugliest things in existence. It cannot be described, but must be seen to be appreciated. We are apt to picture the whole of central Africa as teeming with long caravans of camels, pursuing their way across the deserts, often going hundreds of miles without water, and bearing immense loads. In fact, climatic conditions limit the camel as well as other animals, and when they enter equatorial Africa they quickly succumb to circumstances to which they are not adapted. So, too, exaggeration has played its part in their endurance. In the first division of their stomach the water extracted from their food and drink is stored to tide over a period when none is to

Fig. 429. — Two-humped, or Bactrian, camel (*Camelus bactrianus*).

be obtained, but it is a good camel and a fast traveler that can go a hundred miles without water; and even then it can carry but a small load.

Before the late war the government imported a number of camels to be used in the deserts and alkali plains of the West, but they were soon neglected and allowed to run wild. They, however, found the circumstances favorable, and increased in numbers, and now are quite abundant in some places. Many have been captured and broken in, and occasionally the traveler sees a train of these animals carrying burdens across the plains, presenting a scene that calls to mind the desert regions of the Old World.

The South American camels differ considerably from their eastern relatives in appearance, though but little in structure. They occur only in the cooler regions of the Cordilleras, ascending higher in the mountains,

Fig. 500. — Alpaca (*Auchenia pacu*).

and thus obtaining a proper climate, as they approach the equator. The number of species is a matter of some doubt. Two distinct species occur in a wild condition (the guanaco and the vicuña), while two different forms — the alpaca and the llama — are domesticated, but are unknown in a wild state. Some regard these latter as descendants from the others, while other naturalists regard them as entirely distinct species. According to these views there may be either two or four species in South America.

The alpaca (possibly the descendant of the vicuña) is the better known, at least by name, of the two; for this is the animal whose hair furnishes the well-known dress goods. Less than fifty years ago a consignment of alpaca wool reached England, but no one would touch it, as they could not work it satisfactorily. At last the problem was solved by Titus Salt, and fame, fortune, and knighthood were his rewards. The alpacas will not thrive outside their native country, although many attempts have been made to introduce them into other seemingly suitable regions. Hence all the supply of their wool comes from Peru.

While the alpaca in this way takes the place of the camel as a producer of textile material, the llama fills its position as a beast of burden. Its importance now is far less than in former years. At the time of the Spanish conquest the llama and the alpaca were the only domesticated animals of the Peruvians. To-day the alpaca has gained in importance, while the mule has largely supplanted the other in carrying silver from the mines. About a hundred and fifty pounds is the average load of a llama, and with this it can make scarcely more than a dozen miles a day on the rough mountainous roads.

There are many species of deer, and few wild animals attract more attention than these usually graceful animals, with their curious horns, which usually occur on the males alone. These differ from the horns of sheep and cows in being solid outgrowths of bone, which are shed each year, to be replaced quickly by a larger and more numerously branched outgrowth. By counting the number of these branches one can approximate pretty closely the age of any buck, but still the rule is no more infallible than in the case of the rattles of a rattlesnake or the rings of growth in a tree. The horns in the sheep and cows, on the other hand, never fall off: they also differ in structure; for the bony outgrowth from the skull is covered with a thinner layer of more elastic material, the ' horn ' of commerce.

The musk-deer of the mountains of central Asia differs from most other deer in the absence of horns in either sex, while the upper canine teeth are developed into great tushes in the male, which are the principal weapons employed in the mating time. This deer is celebrated from the musk which it secretes. In the males there is a small sac on the belly into which the secretion is poured, and which may hold nearly an ounce of it. The hunter cuts this out, dries it, and then sends it to the market. Strong as it is in the dry condition, it cannot compare with it when fresh, and it is said that sometimes the hunters are nauseated or even overpowered with the scent at certain seasons of the year. The musk-deer is very shy, and is as agile as a chamois, and hence its capture is a matter of considerable difficulty, a fact which accounts in part for the high price of its malodorous product.

In the reindeer and caribou horns are developed in both sexes. The reindeer belongs to the northern parts of the Old World, the two species of caribou to the same portions of the western hemisphere. These are the woodland and the barren ground species, the differences in habitat being indicated in the common names, the latter being the more northern, while the other just enters our northern states. Both go in large herds composed of a buck and his harem. In the winter they go farther south than they do in summer, and even in the small island of Newfoundland they have their regular annual migrations from one side of the island to the

FIG. 501. — Musk-deer (*Moschus moschiferus*).

other. Caribou-hunting is ranked among the best of sport. The time for it is in the winter, and the amount of suffering which is endured in the chase can better be imagined than described. Sometimes several days will be occupied in the capture of one animal, the whole party bivouacking in the snow for the night, and then following up the trail on snowshoes the next day. The more ignoble trapping and snaring are never adopted by the true sportsman, but for the Indian and the hunter, with whom it is a question of food and clothing, all methods of securing the game are permissible.

The reindeer is the first-cousin to the caribou, and resembles quite closely the woodland species. It is spread across the northern part of the Old World from Scandinavia to Kamtchatka, and in all this long stretch

it is the most important domesticated animal. It is the beast of burden,
dragging the sledges of the Lapps over the snow, and carrying the
Kamtchadale astride its back. It is the chief source of food; for not only
are they milked like our domestic cattle, but their flesh is eaten, and then
the hides, horns, and bones are put to a thousand and one uses. In Lapland herds of five hundred reindeer will sometimes be owned by a single

Fig. 502. — Reindeer (*Rangifer tarandus*).

person; while further to the east, where the animals are more abundant,
herds ten times as large are sometimes seen. The milking is a rather serious
undertaking, for they greatly resent any familiarity with the food of their
offspring. The deer is first caught with a noose as it stands in the yard,
and fastened to a tree or held by some person; but even this does not
always keep it quiet, as it not unfrequently kicks over the milker, and

drags the one holding it all around the enclosure. It gives but half a pint of very rich and strong-smelling milk.

Besides the tame reindeer, the same region is occupied with large herds of wild ones, which may contain two or three hundred individuals. These are hunted for food, and also to replenish the domesticated breeds. Besides, the Lapps allow their own herds considerable liberty at the

Fig. 563. — Virginia deer (*Cervus virginianus*).

breeding season, in the hopes that wild blood may be infused, a matter of some importance, for domestication tends to render the reindeer smaller in size.

Possibly the best known of all our American deer is the species known as the Virginia deer, a rather inappropriate name, as the range of the species is over the whole United States east of the Mississippi River. It is a pretty, graceful form, of a chestnut-red above, and a cinnamon color below, varying to a bluish gray in winter. It is now scarce in many parts where it was once abundant, for it and man cannot well agree; man is too fond of saddles of venison to allow it to remain unmolested. Still, there are many places left where the hunter can engage in deer-stalking, or supply his larder in some less noble way.

In the northern territory west of the Mississippi, the place of the Virginia deer is taken by the white-tailed species, and farther south by the Sonora, or Mexican, deer. The same northern territory is shared by the mule-deer, which receives its name from its long ears. All of these deer have much the same habits, except that one is fond of the woods; another, of the plains; while a third may prefer a mountainous country. All are very curious, and any strange object will quickly attract their attention, and a desire for investigation frequently brings them within range of the hunter's gun. Any sort of vegetable matter will serve them as food; but they are especially fond of rich herbage, and the leaves and twigs of various trees and saplings like the maple.

In an edition of Pliny's Natural History of the year 1601 we read: " More over in the Island Scandinavia there is a beest called Machlis, not much unlike to the alce abovenamed; common he is there, and much talk we have heard of him, how beit in these parts hee was never seene. Hee resembleth, I say, the alce but that hee hath neither joint in the hough, nor pasterns in his hind-legs : and therefore he never lieth downe, but sleepeth leaning to a tree And therefore the hunters that lie in await for these beests, cut downe the tree whiles they are asleepe, and so take them ; other wise they should never bee taken, so swift of foot they are, that it is wonderfull. Their upper lip is exceeding great, and therefore as they graze and feed, they goe retrograde, least if they were passant forward, they should fold double that lip under their muzzle." Such was the olden idea of the animal known in Europe as the elk, and with us as the moose. The old account contains three truths. These animals do live in Scandinavia, they are fleet of foot, and their upper lip is enormous, as our portrait shows. In almost every other respect the picture is a fancy sketch ; the statement about the sleep in a standing position has been made of many other forms, the elephant and the giraffe among others. It is as untrue of one as of the other. The moose is not confined to the " Island Scandinavia." It occurs in Russia as well, and there is a preserve in Prussia. In our own country, as well, the moose is well known, extending its range across the continent, and just entering our northern states.

The moose is very large, and may reach a weight of a thousand pounds. It wanders solitary or in herds of from six to ten through the forest, browsing on the willow and on ' moose-wood ' (striped maple), making on the average a journey of five or six miles a day or better in a night.

When day approaches, they retrace their steps a short distance, and then take a leap from the path, to hide for the day in some clump of bushes. Besides his trail on the ground, the hunter follows the marks of his teeth upon the trees, knowing that he gnaws only on the side which he first approached. Again, another way of hunting them is by imitating the low-

Fig. 504. — Moose (*Alces malchis*).

ing of the male by modifying the voice with a trumpet of birch bark. As the male hears the challenge, he draws near, making his reply, and then when within a certain distance, this call must be stopped, and if one then makes a noise with the paddle to imitate a moose splashing in the water, the game is pretty sure to show himself. Still, the hunt is not over; for a wounded moose, or a male in the rutting season, is not afraid of man, but often forces the hunters to take to the trees.

Civilization has driven the wapati from all the eastern states, and now this large deer is only found west of the Mississippi and in the British possessions. It is much like the red deer, which is so carefully preserved in the forests of Europe for game, and like it may occasionally turn the tables on the hunter, and charge with its sharp, branching antlers.

Allied to these is the giraffe, with its long neck, spindle legs, protuberant eyes, and curious horns, the whole making a strange figure, capable of browsing on the leaves and branches of trees, eighteen or twenty feet above the ground. There is but a single species of giraffe, and this is distributed over tropical Africa, from Abyssinia south to the Orange and Zambesi rivers. It frequents, as a rule, shrubbery not so high as itself, so that while it can readily browse on the tops it can keep a lookout in every direction for any approaching danger. When alarmed, the giraffe runs awkwardly, the legs of the same side moving simultaneously, and not alternately, like those of a horse, and its long neck, stretched forward, swaying from side to side. Although this gait seems extremely clumsy, it is very rapid, and it requires a good horse to keep up with a giraffe. The long neck requires a word. In the birds, length of neck usually means an increase in the number of bones or vertebræ in that region; not so in mammals. Here, with but two or three exceptions, the number of neck-bones is constant. In only one animal does the number exceed seven (in the three-toed sloth), and in the long-necked giraffe and the almost neck-less whales the number is exactly the same, the difference in length being accomplished by an elongation or shortening of the bones.

Between our domesticated cattle on the one hand, and the sheep on the other, occurs a long series of forms differing considerably in appearance, and yet so intergrading that the naturalist has great difficulty in separating them in a perfectly satisfactory manner. Other questions also arise besides this one of classification, and which are no less difficult of solution. Among these is the question of the origin of our cattle; for it is beyond a doubt that it came from some originally wild form. The remains show that originally there were four distinct or easily separable forms of oxen in Europe, the last of which became extinct about two centuries and a half ago, and the probability is that more than one of these was concerned in the origin of the cattle of our fields and barnyards, but that crossing of breeds and artificial selection has produced the great variety of forms recognized by our farmers and stockmen.

Besides these four species there are many other wild forms of cattle, which can have had nothing to do with our tame stock, although some of them have been domesticated. To the American, the bison, or buffalo, of our country is the most prominent of these. In former times the buffalo

ranged over the whole United States, from the Atlantic to the Rocky
Mountains; but as the white man spread west, the herds disappeared
before them, until now there are none east of the Mississippi, and only
two wild herds west of that river, — one in the Yellowstone Park, the
other in British America. None of our animals have been subjected

Fig. 505. — American buffalo (*Bison americanus*).

to such persecution as this. Buffalo robes were demanded by the east,
and the prices they brought led to an enormous slaughter, the body.
merely divested of its hide, being left to decay on the prairies. Then.
too. sportsmen killed them for mere sport. and Indians, for food. The
result is that now these animals. which once ranged over so wide a territory. are now all but extinct.

The buffaloes are gregarious animals. going in small herds, and these uniting in their southerly migrations form the immense bands which formerly were such an element in western scenery in July and August. As they moved onward in their immense numbers, they followed beaten paths, swerving neither to the right nor the left for any ordinary obstacle, climbing hills, and swimming rivers until the spot chosen for the winter was reached. Here they scattered and remained until spring, when they made their way north in smaller parties. On these migrations they were followed by bands of Indians and multitudes of wolves and coyotes, the latter on the constant watch for any weak or disabled one unable to keep up with the herd. These they followed and worried until they were thoroughly exhausted, and then the ravenous horde made quick work with the unfortunate. A noticeable feature in the buffalo is its fondness for water. It enjoys wallowing in a stream or pond; but if these are absent, the herd will take any depression in the ground and roll about in it until they make quite a good-sized hollow, the bottom of which is hard-packed earth. The rain collects in these buffalo wallows, and the herd enjoy their mud-bath immensely. By continued use these wallows become sometimes very large, and form a conspicuous feature in the landscape of the plains.

The European bison, or auroch, was formerly spread over middle Europe, but it is now on the road to extinction. A few remain wild in the Caucasus, while there are about eight hundred preserved in a half-domesticated state in the forest of Bialowieza in Russia.

In India and south into the Malay Archipelago are a number of cattle which are domesticated and put to much the same use as our oxen. One of these is the zebu, with its curious hump upon the shoulders, celebrated as the sacred cattle of the Hindus. The bulls are the most esteemed, and individuals are kept in the temples and treated with a kindness which would be bestowed on no member of the human race, and regarded with a veneration which almost surpasses belief. The banteng, farther to the south, is also domesticated, but is put to more menial duties; while the gayal is remarkable for its enormous horns. The yak lives on the high plains of Tibet; it is another of the forms like our common cattle, but is noticeable for its long hair, which hangs down nearly or quite to the ground from its legs and flanks. In its wild condition it is much like the American buffalo in its habits, while domesticated it is stated to be nearly as valuable and as tractable as our cattle.

In the east, also, the buffalo is also tamed. This is an animal which differs considerably from the buffalo of our western plains, and this confusion of names is a good illustration of the necessity of a scientific terminology. In the strict meaning of the word our animal is not a buffalo, but apparently this name has come to stay. Neither does the employment of the name bison help us materially ; for the true bison is not our species, but the European form mentioned above. The whole case is involved in a confusion which promises to last for a considerable time. The buffalo of the East Indies occurs in a wild state as well as in domestication. Tame, it is a valuable servant, working well and furnishing a small quantity of rather strong-smelling milk ; but wild, it is one of the game animals. Its temper is uncertain, while its strength and courage are such as to have

FIG. 506.—Zebu, or holy cow of India (*Bibos indicus*).

led the Indian poets to compare its onslaught to that of the tiger. When wounded it will charge again and again with great ferocity, and if the hunter be not well mounted, or if he have no tree to shelter him, he is apt to suffer for his temerity. The African buffaloes are no less formidable antagonists, and every traveler's diary contains exciting accounts of struggles with this game. The hides of both make excellent leather, and so assiduously have the African species been pursued that one has now to go many miles north from the Cape of Good Hope before falling in with good buffalo-hunting, while fifty years ago there were large numbers within a few miles of Cape Town.

The line between the buffaloes and a long series of South African forms — the eland, the kudu, the nilgau, etc. — is not a sharp one, while that between these and the oryx, the gemsbok, and the like is scarcely more firmly drawn. All are large animals, quick of sight and hearing and swift of foot, characters which make them favorites with the sportsman. Of these forms there are some forty or fifty species, so that mere enumeration of their names would take more space than we can afford. We can only give room to an extract from Holub's account of the springbok. "The gracefulness of its movements when at play, or when startled into flight, is not adequately to be described, and it might almost seem as if the agile creature were seeking to divert the evil purposes of a pursuer by the very coquetry of its antics. Unfortunately, however, sportsmen are proof against any charms of this sort; and under the ruthless hands of the Dutch farmers, and the unsparing attacks of the natives, it is an animal that is every day becoming more and more rare. The bounds of the springbok may perhaps be best compared to the jerks of a machine set in motion by watchsprings. It will allow any dog except a greyhound to approach it within quite a moderate distance; it will gaze as if entirely unconcerned, while the dog yelps and howls, apparently waiting for the scene to come to an end, when all at once it will spring with a spasmodic leap into the air, and alighting for a moment on the ground six feet away will leap up again, repeating the movement like an India-rubber ball, bounding and rebounding from the earth. Coming to a standstill it will wait awhile for the dog to come close again; but ere long it recommences its springing bounds, and extricates itself once more from the presence of danger. And so, in alternate periods of repose and activity, the chase goes on, till the antelope, wearied out as it were by the sport, makes off completely, and becomes a mere speck on the distant plain."

The term antelope is applied to the forms mentioned since speaking of the buffaloes, and to several to follow. It is a collective term applied to animals with a more or less close structural resemblance to our domesticated cattle, and an external resemblance to deer. Still it is not capable of exact definition. There is but one American species, the prong-horn, or cabree, of the Rocky Mountain region. The most remarkable feature connected with this animal is the fact, first pointed out by Dr. Canfield, that these animals shed the horny covering of their horns, thus differing, so far as is known, from all their relatives. In the deer the whole horn is shed; but in the prong-horn only the outer portion is dropped, the bony core persisting.

The prong-horn is larger than a sheep, yellowish brown above, and white below. It lives in the mountainous country, and depends almost solely on its sense of sight — not upon that of smell — for warning of danger. And when alarmed, what time it will make! there is no American form which can equal it in a race. The bucks are the most watchful and most suspicious, and their warning note quickly alarms the whole herd. Their suspicious-

Fig. 507. — Prong-horn, or cabree (*Antilocapra americana*).

ness is, however, moderated by an undue amount of curiosity. Any strange object attracts their attention: a fluttering rag, the heels of a hunter lying on his back, or any other unfamiliar sight will draw them within gunshot. Dr. Williston relates an amusing instance of this curiosity. "Starting a small herd one evening, and not knowing how useless the attempt would be to follow them, I set out in pursuit. On seeing them go over the brow of a neighboring hill, I crept cautiously for several hundred yards, till I reached its summit. Then rising to my feet, I was myself startled by a shrill snort immediately behind me, and turning about perceived the animals gazing at me in intense astonishment, but eight or ten rods away. They must have followed me for nearly a quarter of a mile, as I crept along in the early dusk."

The prong-horn is easily tamed, and then shows a great fondness for its master or mistress. Even in the wild state they will herd for weeks with the cattle of the settler; while the young are utterly lacking in every

Fig. 508. — Gnu, or horned horse (*Catoblephas gnu*).

particle of fear of man. They are very playful, gambolling about in the most graceful manner, butting each other and their mother, or playing the same game with some bush or tuft of grass. Were the animals better adapted for a life in confinement, they would form a valuable addition to our live-stock; but the efforts of Judge Caton, who has tried the experiment of acclimatization of these as well as many other animals, show that they do not live well in captivity. They do not breed, and they are subject to scrofulous and other diseases which usually carry them off in less than a year.

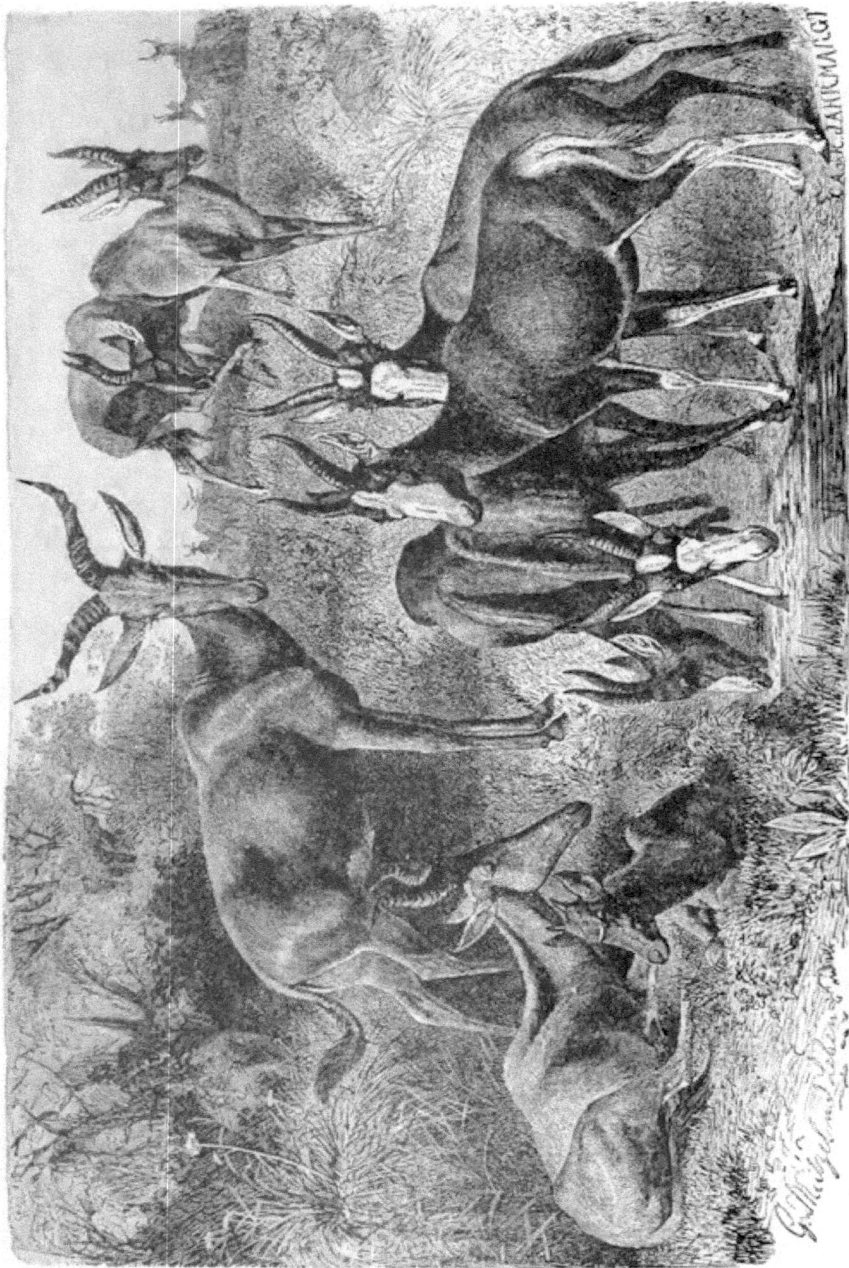

A GROUP OF AFRICAN ANTELOPES.

The harte-beest, the bonte-bok, the bless-bok, the tetel, and its cousin the antelope of Senegal (all but the first figured on our plate) are all African antelopes, which are frequently seen in menageries, and which also figure in every volume of African travel. In habits they are very similar to all the rest, and may be dismissed without further mention. Little also needs to be said of the two species of gnus, otherwise known as wilde-beestes and

Fig. 509. — Chamois (*Rupicapra*).

horned horses of South Africa, one of which is shown in our cut. It needs but a glance at the cloven foot to show that these species are not horses, but are merely strange antelopes. They live in moderate-sized herds, and are pursued by the African colonists for the sake of their hides.

The transition from the antelopes to the goats is furnished by the chamois of the mountains of southern Europe, whose curved and tapering horns are familiar objects on the handles of alpenstocks. The chamois is small, about two feet high at the shoulder, and with a body three feet long. It goes in small herds of twenty-five or thirty, or even fifty individuals, jumping about on the precipitous mountain sides, and climbing cliffs where no less sure-footed animal could possibly follow them. These herds are composed solely of females and young, the bucks leading a solitary life except at the pairing season. When they feed, a sentinel is posted, as is the case with many other animals, and the signal of alarm is a whistling note accompanied with a stamping of the foot. In the Rocky Mountains the mountain goat takes the place of the chamois, living the same shy life in the wildest and most precipitous parts of the mountains, and being almost utterly neglected by the hunters, on account of the worthlessness of the flesh.

Of the domesticated sheep there are many varieties, almost as many as there are of horses or cattle, and in their case we have the same difficulty in tracing their origin as in the case of any other form. The probability is that several distinct species have been mingled in forming the wool-producers of our fields, but this hybridity is not sufficient to account for all the varieties: selection and careful breeding at the hands of man is also responsible for part of the differences between the various forms. One can scarcely imagine the amount of care devoted by breeders to the quality of the stock. "In Saxony the importance of selection in regard to merino sheep is so fully recognized, that men follow it as a trade; the sheep are placed on a table and are studied, like a picture by a connoisseur; this is done three times at intervals of months, and the sheep are each time marked and classed, so that the very best may ultimately be selected for breeding." Of all the varieties thus produced none is more wonderful than the big-tailed Turkish sheep, the tail of which becomes so fat and large that it is a severe task for the animal to carry it about.

Of the wild sheep we mention but one species, the big-horn of the Rocky Mountains, — and this can have had nothing to do with the tame varieties, all of which are derived from Old-World forms. In many of the older works astonishing tales are told of the big-horn, but none more wonderful than that it precipitates itself headlong from the cliffs, strikes upon its horns, which act like springs, and then rebounds to alight all right upon its feet. It is one of the faults of science that it explodes many a pretty tale of travelers, and this one has had the fate of many others: science has shown it to be utterly false. It is difficult to explain how such a belief could have arisen, unless, perchance, it was manufactured from whole cloth. It may be, however, that it grew from an attempted explanation of the battered condition frequently seen in the horns, but even then some of the accounts of "eye-witnesses" must have been deliberate lies. The damage to the horns is explained in a more satisfactory manner. It is caused by the combats of the males in the breeding season. The big-horn ranges from Alaska to California and Arizona and east to the valley of the Yellowstone and the Mauvaise Terres of Dakota. It lives in small herds, the females and young together, while the males go in smaller flocks by themselves. The flesh is highly esteemed.

Fig. 510. — Big-horn (*Ovis montana*).

The ibexes and goats — near relatives — are inhabitants of the mountain regions of Europe and Asia; and in the case of the domestic goat, whose omnivorous tastes are made the subject of many a newspaper jest, we meet the same difficulty which has often been mentioned in these pages, — we do not certainly know from what wild stock it sprang. Some think the original stock of Georgian or Persian origin, while others would seek it in a species occurring in the mountains of Cashmere and Tibet. The economic uses of goats need not be enlarged upon; from their long hair are woven the celebrated shawls of Cashmere, while their hides furnish us with the 'kid' and Morocco leathers, and their milk is made into a cheese.

the peculiar flavor of which is highly esteemed by epicures.   In the ibexes the horns acquire an enormous development in proportion to the size of the head, and the long hair and the 'goatee,' so characteristic of the goats, is lacking.

Last of the ungulates to be mentioned is the curious musk-ox of the Arctic regions of America, but rarely, if ever, occurring south of the sixtieth parallel of latitude, but ranging north to the land of perpetual ice. It does not, however, occur now in Greenland, but it was once abundant

FIG. 511. — Musk-ox (Ovibos moschatus).

there, as well as in Siberia and in northern Europe.   In these northern regions it is a most important animal to the Eskimo.   From its hide, with its long hair, he manufactures his clothing, while his palate does not stick at the strong musky flavor of the flesh of the males at the pairing season. — a flavor which white men, unless starving, cannot stand.   The cows, however, have none of this taste, but are very palatable.   The musk-oxen go in large herds, feeding like the reindeer and caribou, upon sedges and leaves in the summer, and on the moss and lichens in the winter.   In size the musk-ox is smaller than a domestic cow, it being a large specimen which will weigh four hundred and fifty pounds.   In spite of their clumsy appearance they are really swift of foot, and their pursuit calls for the best efforts of the Eskimo. In appearance, as our figure and their name would suggest, they are much like oxen, but in structure they are more like the sheep and goats, and hence are placed near those animals in any systematic arrangement.

WHALES.

" The Name of Whale is given to two sorts of Fish ; one is small, furnished with Teeth, and his Brain produces that white Substance called Parmacity, so much esteemed by the Ladies. The other is the large Whale, who is destitute of Teeth, but then he is supplied with two large Tushes, a dozen or fifteen Feet long, which rise out of his Jaws, and conveniently to amass together the Weeds, which are generally supposed to be his Food, because Quantities of them have been found in his Stomach. These Tushes, split into small Divisions, are the pretended Whalebone, or that strong and pliant Substance we buy of the Merchant under that Name ; and the whole present usefulness seems almost confined to the Hoop-Petticoat : a Mode of Dress altogether senseless and unamiable, but which the Ladies have taken a Resolution to continue, because it gives them less Constraint than the Dress they have now disused." Thus discourseth the ' Count.' concerning whales in that quaint old work. ' Le Spectacle de la Nature,' the design of which was to make scientific facts more palatable by presenting them in conversational form. Yet, while we can readily understand exactly what the count meant, exceptions may be taken to all his statements save, possibly, that relating to the hoop-petticoat, and now even that must not be excepted, for steel has completely replaced the use of whalebone in the manufacture of such articles.

Whales are the most aberrant of all the mammals. They are very fish-like in shape and appearance, and in the older works on natural history they are invariably spoken of as fishes, and even to-day the term ' whale-fishery ' is in vogue, and will probably only disappear with the pursuit itself. But notwithstanding this external fish-like appearance, a whale is indisputably a mammal. In every feature of structure and development it clearly shows this point. It brings forth its young alive, and nourishes them with milk. The skeleton and viscera too are mammalian, and even those features which at first sight seem so fish-like are seen on a closer study to be nothing of the sort. In both fishes and whales the body terminates in a broad caudal fin, but this fin in the fishes is vertical, in the mammals horizontal. In the fishes it is supported by a large number of fin-rays or cartilaginous supports ; in the whale there is absolutely nothing of the kind. Then, too, somewhat similar differences are seen in the anterior appendages. In the fishes the pectoral fins (see p. 304) have a very rudimentary skeleton, while the flippers of the whales have essentially the same bones as are found in the arm and hand of man. Further, whales breathe by lungs; fishes, by gills. Whence the whales had their origin is a far from settled question; but the probabilities are, as shown by Mr. Flower, that they have descended from some ungulated form, and hence we place them after that group. In classification of the whales we follow the old count already quoted, and divide them into two groups, in one of which teeth varying in number are always present, while in the other there are no teeth, but instead the plates of whalebone, the structure and purposes of which will appear in the sequel. In each group there are many species, but of course we can mention but few of them.

The first to be taken up are the toothed whales, — these first because they show the least variation from other mammals, — and of these we begin with the river-dolphins, which occur in the large rivers of India and of South America. They are well provided with teeth (from fifty to seventy in each jaw), which they use in capturing and eating fish, shrimps, and other animals living in the rivers. The species occurring in India are called susu; that in the Amazon, inia. The natives of Brazil invest the inia with certain sacred features, and do not willingly kill it.

Of the family of porpoises and dolphins there are many species, some of which are known by these names, while others have their own special appellatives, as will appear farther on. With us the best-known form is the common porpoise abundant on our shores; and scarcely less common is the dolphin, which is continually coming to the surface, and then diving beneath the waves. They have comparatively little importance except as they devour large quantities of fish. A little less common are the black-fish, which go in large schools, and at times are drawn on the shores in immense numbers by the fishermen. This is especially the case on Cape Cod, and there the annual catch adds a pretty penny to the pockets of the fishermen; for the oil from one fish will average say twenty dollars, and some have been known to yield ten barrels of oil, while small lean individuals will not try out more than that number of gallons. After the blubber has been taken off, the heads are severed from the body, and the oil in the jaws is allowed to run out under the influence of the sun. This is the black-fish oil which is so highly esteemed as a lubricant, and which is used on delicate machinery, watches, and the like.

The black-fish live in constant fear of the killer whales, but these, the most savage of all the whales, do not confine their attacks to the black-fish. Anything that swims is their prey. They attack even the largest whales. "The attacks of these wolves of the ocean upon their gigantic prey," says Captain Scammon, "may be likened in some respects to a pack of hounds, holding the stricken deer at bay. They cluster about the animal's head, some of their number breaching over it, while others seize it by the lips and haul the bleeding monster under water; and when captured, should the mouth be open, they eat out its tongue." The same author describes the attack of three killers on a California gray whale and her calf which had grown to three times the bulk of the largest killer. The contest lasted for an hour or more. "They made alternate assaults upon the old whale and her offspring, finally killing the latter, which sank to the bottom where the water was five fathoms deep. During the struggle the mother became nearly exhausted, having received several deep wounds about the throat and lips. As soon as the prize had settled to the bottom, the three killers descended, bringing up large pieces of flesh in their mouths, which they devoured after coming to the surface. While gorging themselves in this wise, the old whale made her escape, leaving a track of gory water behind." The killers are not afraid of man, and instances are known where whalers have had their prey forcibly stolen away by these animals while towing it to the ship.

The white whale rarely appears south of the mouth of the St. Lawrence, but there it was once very abundant, and it is from this locality that the specimens occasionally exhibited in aquaria are obtained. In confinement they are fed on fish, chiefly eels, and of these it takes between two and three bushels to supply the daily needs. They have been fished to death for their oil and skins. The latter makes an excellent leather, soft and very durable, and almost impervious to water. This whale-fishing has had a result which is very interesting, as showing the interdependence of nature. Since these whales were reduced in number it has been noticed that the number of salmon in the Canadian rivers has greatly decreased, and to an extent not to be explained by any over-fishing. The chain seems to be as follows: the white whales feed on seals, while seals in turn are very fond of salmon; the decrease in whales of course resulted in an increase of seals, and this in turn had its influence on the salmon. Now that the seals are persecuted and killed by the hundreds of thousands by man, the salmon-fishery is improving. It is almost a case of Tenterden steeple and Goodwin sands.

The strangest of all the whales is the narwal of the Arctic seas, an animal from ten to fourteen feet in length, and armed in front with a long, spirally twisted horn of ivory, which may add from six to ten feet more. This long tusk, which sticks straight forward, is but a greatly modified and enormously developed tooth, and is the only dental armature existing in the adult creature, which is said to feed on fish and soft marine animals.

Most celebrated of all the whales is the sperm-whale, one of the largest of all mammals, which occurs in all seas except those of the extreme north. In length it may reach eighty feet or even more, and a glance at our cut will show what a proportion of this length is occupied by the large, square head, which has an outline varying strangely from that of the skull. Between skull and skin in the front part of the head is an

Fig. 512. — Sperm-whale (*Physeter macrocephalus*).

enormous deposit of flesh and fat, and on the right side is a large cavity — the 'case' of the whalers — which is filled with oil and spermaceti, sometimes to the amount of fifteen barrels or more. On the capture of one of these whales the first thing is to cut into this case and take out the oil and wax in buckets.

The sperm-whale has teeth, forty-eight in number, in the lower jaw alone, and these fit into a groove in the upper jaw. These teeth, of the whitest ivory, may be four, or even six, inches in length. These are of great use in eating the squid and fish, which form the food of these huge beasts.

They swim through the water with the lower jaw let down (as shown in the cut), until it forms a nearly right angle with the body, and thus sweep into the mouth all sorts of animals, even the giant squid; already noticed on page 122 as food of this species. These are bitten and torn by the teeth into morsels of a suitable size, which are passed to the capacious throat, large enough to allow the passage of a man.

For the greater portion of our knowledge of the habits of this, as of almost all other whales, we are indebted to the pages of Captain Scammon, which have been freely used in the preparation of this article, though credit is not always given by quotation marks. No whale respires as regularly as the sperm-whale. On coming to the surface the hump of the back first appears, and then the head, which sends forth a volume of whitish vapor like steam. This is the spout, and may be seen from the mast-head at a distance of three to five miles. The largest bulls take ten or twelve seconds for an inspiration, and will blow sixty or seventy-five times at a rising; when the 'spoutings are out' they pitch downward, head-foremost, turning the flukes high in air, and descending in a perpendicular direction to the great depths, where they remain from fifteen minutes to an hour and a quarter. When spouting undisturbed the whale lies still, or moves along slowly through the water.

When alarmed, or when sporting in the water (for whales like their play), the actions are different. If frightened, they may instantly sink, or if startled, they frequently assume a perpendicular position, with the greater portion of the head above the water, so as to more readily look and listen, or, lying horizontally, they will strike right and left with their powerful tail, to see if any object be within reach. When at play, the flukes may be raised high in air and pounded upon the water, raising a cloud of foam and spray; or, descending a few feet, it will dart suddenly upwards, at an angle of about forty-five degrees, until nearly the whole body is out of water, when it falls on its side with a heavy splash, which may be seen for miles. When attacked, the sperm-whale makes a desperate fight, and not infrequently succeeds in escaping; but occasionally, after being struck, it will lie motionless for a few moments, allowing the lancer to get in his murderous thrusts. At other times it takes the offensive, and crushes the boats with its jaws, or, coming head-first, strikes the ship again and again, with such force as sometimes to cause it to sink.

Floating on the surface of the sea are frequently found large masses of an odorous substance known as ambergris, and which always commands a high price. Many are the theories advanced to account for its origin, some extremely fanciful and absurd. We now know that it is produced only by diseased whales, and consists of the excretions which have been long detained. Its peculiar scent is due to its being composed of squid, and the jaws of these animals are frequently found in it. Douglass, writing a hundred years ago, says, "squid-fish, one of the Newfoundland baits for cod, are sometimes in Newfoundland cast ashore in quantities, and as they corrupt and fry in the sun, they become a jelly or substance of an ambergrease smell; therefore as squid bills are sometimes found in the lumps of ambergrease, it may be inferred, that ambergrease is some of the excrement from squid-food, with some singular circumstances or dispositions that procure this quality, seldom occurring." Masses of ambergris weighing two hundred and twenty-five pounds have been obtained.

The two-toothed whales, most of which are inhabitants of the Southern Seas, and which have only two teeth in the lower jaw and none in the upper, can only be mentioned by name. A relative of these is the bottlenosed whale of the northern Atlantic, in which it requires no little imagination to see the slightest resemblance to a bottle in the rounded head and the protuberant jaws. The oil of this species is nearly equal in quality to that of the sperm-whale. The bottle-nose is destitute of teeth.

The whalebone whales which follow are utterly destitute of teeth in the adult condition, and yet the presence of teeth in the young is conclusive evidence that these animals have descended from toothed forms. In place of teeth they have a most curious strainer composed of large horny plates placed one after another on either side of the upper jaw, and fringed out below into large numbers of rather stiff hairs. These plates furnish the whalebone of commerce; they are not true bone, but are more of the nature of hair agglutinated together. When a whalebone whale opens its mouth, the plates of whalebone or baleen drop down, so that they form a curtain extending on either side from one jaw to the other, and then when the mouth shuts again, the plates are carried up into a hollow in front of the throat. All sorts of marine invertebrates form the food of these whales, and the minute forms that throng the seas are known to the sailors as 'brit.' The whalebone whales swim open-mouthed through the water, their jaws and the whalebone forming a fine net which retains the tiniest particles, and on these, insignificant as they would seem, the body, sometimes a hundred feet in length, is nourished. Were the sea as barren of these forms as is fresh water, the whalebone whale would quickly starve; but the sea absolutely teems with microscopic life, and a few minutes' work with the surface net of the naturalist will give one new ideas of the sufficiency of these infinitesimal animals to support the largest forms of life.

We need say but little of the species of whalebone whales; indeed, naturalists have scarcely gotten them straightened out. Still there are some species which are perfectly well known. Of these is the devil-fish, or gray whale of the Pacific, a form which has a large number of other names, and which was formerly very abundant on the California shores, a thousand a day passing southward in winter within sight of land; to-day the average seen from the same stations would not exceed forty. Shore fisheries are established at various points, which have reduced the numbers. The hump-backs are larger forms occurring in both oceans; but larger than these are the sulphur-bottoms of the Pacific, the largest of all animals existing in the world. In Scammon's pages are given the dimensions of one animal: length, ninety-five feet; girth, thirty-nine feet; length of jawbone, twenty-one feet; yield of whalebone, eight hundred pounds; yield of oil, one hundred and ten barrels; weight of whole animal (by calculation), two hundred and ninety-four thousand pounds. It is not impossible that individuals a hundred feet in length occur; but whalers neglect this species for very valid reasons. It is the strongest of all whales, and swims with remarkable rapidity, so that until the introduction of the bomb-lance, it was almost impossible to capture it. Then, too, its oil is not so good as that of many other easily taken species.

In the Atlantic it is replaced by the silver-bottom and the fin-backs, the latter but slightly inferior to the sulphur-bottom in size. Of late years, since the almost total cessation of the whale-fishery, they have become very abundant in the Gulf of Maine, and I have seen ten or a dozen in sight at one time. One has no idea of their immense size until one rises a few fathoms away from the boat, blows, and then goes down headforemost. They are very active, and as they sink when killed, their pursuit is accomplished with much difficulty. Within recent years their fishing has been prosecuted by means of steamers from Provincetown and Boothbay, but it has not been very successful from the financial standpoint.

The bow-head, though much smaller than these species just mentioned (it rarely exceeds fifty feet in length), is by far the most important of all the whalebone whales, on account of the enormous amount of oil and bone it yields. Some of the plates of whalebone may be a foot wide and fourteen feet long, and of this bone, long and short, over seventeen hundred pounds has been taken from a single individual, while another has yielded two hundred and seventy-five barrels of oil.

The bow-head is more Arctic in its habits than the right-whales, of which there are no one knows how many species. In olden times they were very abundant on our eastern coast. Then, owing to fishing, they almost entirely disappeared, and now they are once more becoming abundant, and a fishery for them has been established in the Carolinas. They are usually confounded with the bow-head, and hence there is some difficulty in separating them, in the accounts of whaling expeditions, from that species. The New Bedford fisheries first exhausted the Atlantic and then turned to the Pacific Ocean, and in 1875 I find records of individuals, taken in the neighborhood of Bering Sea, yielding two hundred and eighty and ninety barrels of oil. The name right-whale means the right kind to kill for whalebone: for these stand second only to the bow-head in this respect, and formerly that species was included by the whalers under the same common name.

FIG. 513. — Razor-back whale (*Balænoptera musculus*).

We have given but a few incidental allusions to the whale-fishery in the preceding pages, because the subject could better be briefly described in one place. Whaling-ships are small vessels of from three to five hundred tons' burden, and carry four whale-boats, and a crew, all told, of thirty-five men. Arrived at the whaling-grounds, a constant lookout is kept at the mast-head, and the welcome cry of "there she blows," "there she breaches," or "there go flukes" sets the whole ship in a flutter of excitement. The boats are quickly lowered, a mate taking charge of each, and with him are four of the strongest men in the ship, the one rowing

the bow-oar being called the 'boat-steerer,' though the mate steers until
the whale is struck. On starting out, everything is properly arranged.
The line is coiled up in the tub, the end extending forward over the oars
to the bow, where it is fastened to a harpoon. Each boat is rowed as fast
as possible; for there is no little rivalry between the various boats' crews
as to which shall first strike the whale. As they draw near, the speed is
slackened, and perfect quiet reigns; for there is danger of alarming the
whale before there is a chance to throw the harpoon. At last the mate
gives the command, the boat-steerer jumps up, and with all his might
darts the harpoon into the body of the monster before him. Now is the
time of excitement. Instantly the command is "stern all," the object
being to get out of reach of the flukes of the animal, which are lashing the water into a foam, and which would break in the sides of a boat
as quickly as if it were made of paper. The mate and the boat-steerer
now change places; for the mate attends to the lancing and killing of
the whale.

The tactics of the whale vary with different individuals: sometimes they
lie still upon being struck, so that the mate can at once dart in the lance
and touch some vital point; sometimes they start straight off, dragging the
line so swiftly that it smokes as it runs through the brass grooves, and
water has to be poured on to keep it from burning. Woe be to any luck-
less individual who should get caught in the coils as they run out. Again
the whale 'sounds,' or descends to great depths, and then the endeavor is
to check the descent as much as possible by grasping it with nippers; but
in this great care must be exercised, or the boat itself will be drawn under
water; then the line has to be cut.

Sooner or later the whale becomes exhausted, and the boats row to it
as it lies upon the surface, and the fatal lance is plunged in and churned
again and again, until finally the huge animal dies. Within recent years
the old-fashioned harpoons have largely given way to murderous bomb-
lances — large cartridges which are thrown by hand or are shot from a gun,
and which explode inside the victim. It is only with these that the sulphur-bottoms and the fin-backs can be successfully pursued. In the case of
the latter, which sink when they die, a harpoon is thrown at the same time
as the bomb, and to this is attached a long line and a float, the object of
which is to show the position of the corpse beneath, which will rise in a
few days, swollen by the gases caused by decomposition — a time anx-
iously waited for by the captain and crew.

After the whale is killed, it is towed to the ship, or the latter sails
to the prize. Then the body is fastened by chains so that it can be
easily handled, and with knives and spades the blubber is taken off and

hauled on board to be tried out. The whole process is very intricate, and in it the whale is turned over and over until at last all the flesh is gone, when the skeleton is left to the sharks which have gathered round. The blubber, which may be nearly a foot in thickness, is placed in the rendering-kettles and tried out, placed in barrels, which are headed up and stored below.

The history of the whale-fishery is an interesting one. It was first prosecuted as a regular occupation by those brave fishermen of the Bay of Biscay, who long before Columbus made their regular voyages to the Banks of Newfoundland in search of cod. From them the business passed to the Dutch, and thence to the inhabitants of the New World, where Nantucket and New Bedford soon became pre-eminent. They cleared the Atlantic of whales. They ventured into the Arctic seas and sailed around the two great capes to the south into the Pacific in their search for oil, and at one time seventy thousand people were directly or indirectly employed in the United States in the pursuit of oil and bone. To-day the whole business has changed, and its former glory has departed, a fact due principally to three things: the decrease in the number of whales, the result of over-fishing; the discovery of the oil-fields of Pennsylvania and the consequent introduction of petroleum for lighting and lubricating; and thirdly, the fear of privateers in the late war to whom a whale-ship was a welcome prize.

SEA-COWS.

Columbus in one of his voyages saw a mermaid; at least, so the story goes. What he really saw, of course, it is not possible to say with certainty, but the chances seem to favor its having been a most peculiar animal, the manatee, or sea-cow, of the tropical Atlantic. If we get a close view of the creature, the resemblance to what a mermaid should be is exceedingly remote; but seen at a distance with but the head above the water, and at a time when the imagination was all excited by the wonders of the New World, it is possible that this creature may have been transmogrified into one of those animals of whose actual existence no one at that time had the slightest doubt.

The manatees are very whale-like in appearance, and cannot be removed far from them in any systematic arrangement, and yet they have many peculiarities which forbid their union with them. But two distinct genera and less than half a dozen species are now living; about a century ago there was one more. The fate of this is very interesting. Bering, the Russian explorer who gave his name to an island in the Arctic seas and to the strait separating America and Asia, visited Bering Island over a century ago. He found on the shores of that one island a huge manatee-like animal from twenty to twenty-eight feet in length, and Steller, the naturalist who accompanied him, gave a description of it. It formed great herds, surrounding the shores, where it fed upon the sea-weed, grinding it between the horny plates which took the place of teeth. It was very tame, even stupidly tame, and this led to its destruction. It furnished while it lasted an abundant supply of fresh meat, and Middendorff supposes that the last individual was killed in 1768. It was called *Rhytina*.

The dugong of the Indo-Pacific Ocean is not yet extinct, though its numbers have greatly diminished in the last two hundred years. In olden

FIG. 514.— Dugong (*Halicore dugong*).

times they might be seen, as around the Mascarenes, feeding in large herds in the sea-weeds in the shallow water, but they were possessed of too many valuable qualities to allow them to remain free from interference by man. In the first place, their flesh is very good, especially that of the calves, while from the fat a clear oil can be tried, which has no unpleasant taste, and which besides will not become rancid in the warmest weather. Around Bourbon and Mauritius they have been largely killed off, while on the northwestern coast of Australia a dugong-fishery has been established, conducted in much the same way as the capture of whales, and this has had its effect in these regions. The dugong has a much wider distribution than the northern sea-cow, and so may escape extermination for a much longer time.

The manatees, three in number, inhabit the Atlantic and its tributaries. One is African, one ranges from the West Indies to Brazil and up the Amazon to Pebas, Ecuador, while the third inhabits the shores of Florida. Of these the best known is the South American form. On the whole course of the Amazon it is fished for by the natives, who use harpoons or nets for the purpose. As soon as caught the nostrils are plugged up, and the creature is allowed to die of suffocation. The natives eat the flesh, and seem to be very fond of it. It tastes like coarse pork, but the fat which occurs in thick layers between the muscles is greenish in color, and has such a fishy taste that only starvation will bring a European to eat it. Like the other sea-cows the manatees feed on the pastures of aquatic plants. They are very stupid, but they seem to exhibit considerable conjugal affection.

THE CARNIVORES.

The word carnivores means flesh-eating, and is eminently appropriate for the long series of cats, dogs, bears, seals, and all their relatives. They are fully fitted for a diet of this sort. They possess ample means for capturing their prey, — strong claws for tearing it in pieces, and teeth which are splendidly adapted for the cutting of muscles, the severing of tendons, and the crushing of bones. Still, they are not all exclusively flesh-eaters: the bears exhibit, as is well known, a fondness for fruit and honey; the raccoons will plunder gardens; even cats occasionally exhibit a sweet tooth, and will eat dates, green corn, and the like; while the honey-bear of the East Indies almost wholly ignores flesh, and makes its meals on honey, white ants, and various fruits. These instances, which might easily be multiplied, are, however, the exceptions; the carnivores are, as a whole, pre-eminently carnivorous. They are divided into two groups: first the typical terrestrial forms, like the dog, bear, and cat; and second, the thoroughly aquatic seals, walruses, and the like.

First among the terrestrial carnivores come the bears and the bear-like forms, the most prominent feature of which is seen in their attitude and mode of walking. A dog or a cat stands on the tips of the toes, the wrists and ankles being raised some distance above the ground. In the bear-like forms, on the other hand, the whole palms and soles, from the wrist and ankle down, are applied to the ground. The first-mentioned group are hence frequently termed digitigrades, or toe-walkers; while the bears, and their relatives, are plantigrades, or walkers on the sole of the foot.

In northern Mexico, and extending some distance into the United States, is a curious little animal which bears no less than a dozen English names, besides a number more of Indian and Spanish origin, but what one of these is best is a question. American civet-cat, Mexican cat, and ring-tailed cat are all objectionable, because they infer a relationship to the civets and the cats which in reality does not exist. Cat-squirrel is hardly better, while raccoon-fox would do were it more popular. The reader may take his choice. This polynomial animal has the appearance of both a cat and a raccoon, and a very pleasing physiognomy. It lives in the trees and among the rocks, and if it can but find a deserted ranch, its home is made in it. It is quiet in its habits, moving about but little until after dark, and then it begins its prowling. It is fond of rats and mice, while a chicken-coop presents attractions and temptations sufficient to overcome a not very strong conscience. In Mexico, as in California, it is frequently caught and tamed, and makes a very interesting and amusing pet. It is very playful, and we can only regret that living specimens have not fallen into the hands of those of our naturalists who are able to write those entertaining animal biographies. It has, however, made the scientific world much trouble; for it has been very difficult to assign it to its proper place in the animal series. *Bassaris* is the name naturalists have given it.

It would, however, appear that the ring-tailed cat is most nearly related to our next form, the raccoon, or coon for short — that nuisance in the cornfield or in the poultry-yard, that delight of the colored citizens of the southern states. Who is there that has not seen a coon-hunt, or at least heard one described? A coon-hunt properly conducted requires negroes; otherwise, it is about as spiritless as trout-fishing in a trout-pond, or gunning for game-birds in a poultry-yard. In a coon-hunt the hunters afford as much sport as the hunted. Besides the negroes are the coon-dogs, lank creatures fully alive to all the pleasures of the nocturnal expedition, and trained to distinguish the scent of a coon, no matter how deep he may be in some hollow tree. As the party starts out lighted with torches and lanterns, some carrying axes, and others shot-guns, the dogs running eagerly along, and all in a high state of excitement and volubility, the scene is beyond description. Of course the dogs and their baying and barking soon inform the party that a coon is treed. Guided by the sound, they make their way across meadows and ditches, through swamp and woods, until at last they reach the tree. If the coon be in the branches, his glistening eyes, lighted up by the torches, are soon seen, and then the hunt is quickly over. It, however, frequently happens that the tree is hollow, and his coonship is hidden deep in the cavity. Then the axes beat against the trunk to see how far down the hollow extends. At the bottom a hole is cut, and then dry leaves and grass are thrust in and set on fire. The smoke is more than the coon can stand, and he soon rushes out of the upper opening, and then falls a victim to the murderous shot-gun. Occasionally he may drop to the ground, and take to his legs, fol-lowed by dogs and negroes, each exercising his lungs to the utmost. When

Fig. 515. — Common raccoons (*Procyon lotor*).

the dogs catch up, they must have a care for themselves: for a coon at bay is no mean opponent. He has sharp claws and strong teeth, and throwing himself on his back, he will fight desperately for life; and frequently will not succumb until hit on the head with a club or the butt of a gun.

To white people, nourished on tame flavored beef and mutton, the meat of the coon is not over-palatable; it has a rank, gamy flavor, which one can hardly relish. To the negro, however, this taste adds a peculiar piquancy to the meal, and chicken and bacon will be neglected when coon can be had.

The coon is about two feet long, with a tail of a foot more. Its color is a gray varied to a dirty yellow, and mottled with hairs tipped with black. These black hairs are most numerous upon the face and in the rings of the tail. The raccoon lives in hollow trees, and in our northern winter it hibernates much as does the bear, passing all the colder months in a torpid condition. While the coon is a carnivore, and is especially fond of birds, small mammals, and fish, he is not averse to an addition of vegetables to his diet. Green corn and pumpkins are possibly preferred, but when one of these animals gets in a garden, it commits sad havoc. It seems to destroy from sheer wantonness; for it pulls down far more than it can possibly eat. It is said that it dips all its flesh into water before eating it — a habit which is exceedingly difficult to explain.

Relatives of the raccoon are the coaitis or coaiti mondis of the warmer parts of America, from Mexico to Brazil and Paraguay. The more southern of the two species is shown in our plate in an admirable manner, and all that is necessary is to say that it is brownish above, and yellow or tawny beneath, while the tail is banded with black and dirty orange. It is the commonest carnivore in all South America, and is remarkably mild in its disposition, even when wild. The Indians catch it by shooting with darts poisoned with curare, and then if they do not wish to use it for food, they cure it by rubbing salt — the antidote to curare — into the wound. The flesh is said to resemble beef, but to have a sweeter and richer flavor. Still this animal makes such an engaging pet, that it is a pity to think of it as an article of diet. Dr. Samuel Lockwood has given a most entertaining sketch of one which he kept in confinement, and from his account we take freely in the following, using largely his own language, for it could not be bettered.

Inappeasably inquisitive, she was incessantly thrusting her long and flexible nose into everything. It was enough that anything was hollow to excite her curiosity. The dinner-bell was turned over; but, unable to detach the clapper and chain, it was soon abandoned in disgust. A round sleigh-bell received more persevering attention. Unable to get her nose or paws into the little hole at the side, the clatter within set her wild with excitement, and evoked a desperate attack on the little annoyance with her teeth. She then gave it up as a bootless job. A bottle of hartshorn was made the next subject of investigation. She was not in the least disconcerted by the drug; in fact, she had a strong nose for such things. Then came a tobacco-box. Resting it on the floor between her two paws, she turned it over and over, using her nose, claws, and teeth upon it with great energy, but to no avail. The smell of the contents seemed to enrapture her, and when the box was opened for her, she seemed in raptures. In went the nose, also both fore paws. Very soon that wonderfully mobile organ had separated every fibre, so that the mass seemed trebly increased. Next came a dirty pipe, and her velvety nose was instantly squeezed into the rank nicotian bowl. The old cat had just finished her nap, and was stretching herself, an operation which means that she stood with her fore feet close together, the limbs elongated, the back arched up like that of a camel, the head drawn back and yawning widely. Such a sight the coaiti had never seen; hence it must be looked into. So in a trice, erect, and resting flatly on her hind feet, like a little bear, she put her arms around Tabbie's neck, and, reeking with nicotine, down went that inquisitive nose into the depths of the feline fauces.

The coaiti is an amiable animal, and even when two are kept together, they never quarrel even over their food. They will eat almost anything; milk, sugar, insects, worms, rats, mice, and poultry all come within the scope of their appetites. Curious is the way in which they search for earthworms. That long and mobile nose, like the snout of the pig, roots up the ground, and ploughs long furrows an inch in depth, through garden and greensward. When attacked, there is no disposition to run, but, instead, a show of courage, no matter what the odds.

In the kinkajou of South America the tail reaches its highest development outside the monkeys. It is very long, and like the same member in the monkeys, it is a fifth hand. It can be coiled around the branches of a tree strongly enough to support the weight of the animal; while the hind feet are almost as useful as the front pair in handling objects of food.

The bears, as a whole, are the least carnivorous of any of the present group. Their general appearance needs but little comment here. In shape all are much alike, and there is comparatively little diversity in their habits. They are the typical plantigrades, and their gait seems heavy and awkward, and yet they are able to get over the ground with considerable rapidity. There are only a few species in the whole world, and North America boasts of but four, together with several geographical varieties. These are the brown, black, grizzly, and polar bears.

Of these the brown bear is a resident of the more northerly portions of our country; but is far more abundant in the Old World, stretching across from Scandinavia to Kamtchatka and Japan, and south into the Pyrenees. It is larger than our black bear, and is a familiar animal in captivity, as specimens can be seen in every menagerie and zoological garden, while it is the most common bear of the strolling mountebanks.

COATI MONDI (*Nasua rufa*).

In the United States its place is taken by the black bear, which can at once be distinguished by the glossy black color of its long, hairy fur. Besides this typical form there are some varieties. Thus in the Rocky Mountains occurs the cinnamon-bear, the fur of which is a dark chestnut or cinnamon color; while in our southern states occurs a yellow bear. These

Fig. 536. — Brown bear (*Ursus arctos*).

are both without much doubt varieties of the common black bear. Of the black bear itself, the hunters recognize two types. One is a long-legged, long-bodied form called the ranger-bear; while the shorter-bodied, shorter-legged, and blacker form is called the hog-bear. In this case it is certain that the differences are those of mere individual variation, as two cubs from the same litter have been known to develop in the two different lines.

In their wild state the bears range through the woods in search of food. They eat berries and beechnuts, acorns, roots, and the like, and even the intensely acrid bulb of the Indian turnip. If a clearing is found, they will eat oats, green corn, and pumpkins. Fish, too, has its charms for them, but they are not especially fond of other meat. Sweets, however, are the greatest luxuries, and honey and molasses will cause bruin to leave any other food. The black bear makes a nest, in some secluded spot, of leaves and bark, and to this it retires when its day's labor is over. In December or January it enters this nest for its long sleep of three or four months. It covers itself completely, the ears excepted, with the leaves, and rolled up in a ball, with the paws crossed over the nose, it sinks into a deep lethargic sleep. During all this long time it takes no food; it does not even suck its paws, as is often asserted. It merely sleeps, breathing slowly, and with slight inspirations and expirations. Every muscle is at rest; there is not the slightest exertion during this long period of torpidity, and hence there is but slight drain on the system and the store of fat. In the spring, when this hibernation is over, the male bear leaves his nest in the rocks, nearly as fat as when he entered it in the fall; but now, even if he find food comparatively abundant, he loses flesh rapidly, and soon becomes very lean. The female, on the other hand, is lean when she comes out from her retreat. The reason for this difference is not far to seek. It is in the winter that her young are born, and the little cubs suckle for two months before the mother leaves the nest, and all this time she is without food, and the milk is formed from her fat stored up in the previous summer. The young, when born, are very small and helpless. They are not six inches long, weigh less than a pound, and are not covered with hair. They are blind, and their eyes do not open for a month.

The principal cause of hibernation seems to be a question of food supply. As long as mast remains abundant, so long will the bears prowl about, no matter how severe the weather may be. Hibernation in the case of the bears seems to be more a provision to bridge over a time of scarcity of food, than any physiological requirement caused by temperature and the like. In the case of the female, however, another element — the approaching maternity — is to be mentioned in this connection.

The black bear is far from ferocious. It will turn and run from a man if it can have a chance, but if cornered it will fight. A female with cubs is, however, a more dangerous beast; for she is always ready to defend her young, and will even take the aggressive. Bears strike the enemy and try to throw him down, and then bite and tear him. It is usually said that they try to hug and squeeze a man to death, but old hunters deny this. They grasp the victim with the fore arms and hold him tight, but this is merely for the purpose of giving the strong claws on the hind feet a chance to tear and lacerate the body, and these organs make even worse wounds than the teeth.

Among savage, and even among some civilized people, the bears are viewed with a peculiar reverence, which, however, does not prevent their being killed whenever opportunity offers. In northern Europe this goes so far that the name of the bear must not be spoken, and as in the case of the Jews of yore, they have invented various euphemistic expressions, such as the ' old man in the fur coat,' the 'dog of God,' and the like, for use when referring to this animal. Our own Indians also recognize a supernatural element in bruin, as in the following account from the pages of Charlevoix : —

" As soon as a bear is killed, the huntsman places his lighted pipe in his mouth and blows the beast's throat and windpipe full of the smoke, at the same time conjuring his spirit to hold no resentment for the insult done his body, and to be propitious to him in his future huntings. But as the spirit makes no answer, the huntsman, to know whether his prayers have been heard, cuts off the membrane under his tongue, which he keeps till his return to the village, when every one throws his own membranes into the fire after many invocations and abundance of ceremony. If these happen to crackle and shrivel up, and it can hardly be otherwise, it is looked upon as a certain sign that the manes of the bears are appeased ; if otherwise, they imagine the departed bears are wroth with them, and that next year's hunting will be unprosperous, at least till some means are found for reconciling them, for they have a remedy for everything." It must be a comical sight to see one of these grim warriors bending over the carcass of bruin and whispering in his ear : " Now, bear, it was a fair fight, and you were beaten, so you must not bear any resentment any more than I should if you had killed me."

The grizzly bear of the Rocky Mountains is the largest of all the group, and well does it deserve the name of *horribilis* if a tithe of the stories told about it are true. It may reach a length of eight or nine feet and a weight of seven or eight hundred pounds, and its strength is in due proportion to its size. Its coat is a grizzly gray or brown, its teeth are strong, and its claws are long and sharp, making it the most formidable opponent of all the bears. Many are the tales of its hunting its hunters and its turning the tables on its pursuers.

The polar bear, the last American species to be mentioned, lives far to the north in the realms of snow and ice, and its fur, of a snowy or yellowish white, agrees with its normal surroundings. In the regions it inhabits vegetation is wanting, and so it must perforce feed almost solely on flesh.

It lives largely on fishes, seals, and the like, and to catch these it must take to the water. It is a strong swimmer, and when after seals it swims quietly with only the nose above the surface until it is within reach of its prey. On the land it pursues the foxes, ptarmigans, and other terrestrial forms. Its large size — slightly less than that of the grizzly — makes it even a formidable opponent to man.

FIG. 517. — Grizzly bear (*Ursus horribilis*).

One or two of the foreign species need to be mentioned. First comes the subject of the adjacent plate, which rejoices in a multiplicity of names, — sloth-bear, large-lipped bear, jungle-bear, honey-bear, aswail, and bear-sloth, — from which the reader can take his choice. It has a most curious physiognomy. At an early age its incisor teeth drop out, and this, together with its enormous lips, its mobile nose, and its strange eyes, make it at once

HONEY-BEAR, SLOTH-BEAR, OR ASWAIL (*Melursus labiatus*).

striking and homely. It is of a retiring disposition, and yet when cornered or excited it is far from a contemptible adversary. It always strikes for the face and eyes with its long and sharp claws. It feeds on wild honey, white ants, and fruit, and when occasion offers, on other substances, but rarely on flesh. It lives in the caves and in the hollow trees of India and Ceylon; it climbs well and prefers not to leave the jungle. In captivity it is mild and about as intelligent as any other bear.

In the same region, and also ranging farther to the east, is the bruang, or sun-bear, a smaller but similarly marked species which makes a really amusing and interesting pet. It displays much affection and is almost invariably good-tempered. In the menagerie it shares the honors with the monkeys, it is so active and so strange in its actions. It will raise itself on its hind legs and bow its body with all the dignity if not the grace of a Chesterfield, and then it will go through a series of facial gymnastics which would put any clown, even George Fox, the famous Humpty Dumpty, to shame. Then its somersaults are things to be seen; they cannot be described, they are so ludicrously clumsy. All of these acts are natural, and do not proceed from any training. The animal is forced to go through them in just the same way as the monkeys are. He cannot help himself, he is so brimful of animal spirits.

The bears form the central group of the bear-like forms. On the one hand, we had the raccoon and the coaiti; on the other side comes a much larger series of forms, of which the weasels, skunks, badgers, and others may be regarded as representatives. It is a very important family reckoned from the economic standpoint; for to some of its members we owe the most valuable of furs, — ermine, sable, martin, mink, sea-otter, etc., — which will be alluded to in their proper place. It is also remarkable for the development of glands near the vent, which secrete a strongly odorous fluid, which in some, as the weasels and martins, seems to be principally of use in calling the sexes together; but which in the skunks, pole-cats, stinking badgers, and the like, becomes a very important means of defence.

The sea-otter of the coast and islands of the northern Pacific is the most peculiar member of the group, fitted as it is for an aquatic life and a diet of fish and shell-fish. It lacks the sharp-edged teeth of the others, and it has its feet webbed, the hinder pair being broad, so that the animal bears no little resemblance to a seal. In color it is a 'deep liver-brown,' the surface being silvered by the longer and stiffer gray hairs, which extend beyond the rest of the pelage. It was first brought to the notice of the scientific world, in 1751, by Steller, the Russian naturalist who accompanied Bering to Kamtchatka and the Bering Sea. It had been known long previous to that time by the natives, and in the early days of Russian possession of those northern shores, almost fabulous numbers were killed, two men killing five thousand on one island in a single year. The skins commanded enormous prices in China, and the result was that their numbers were soon decimated.

Saanach Island is now the great sea-otter hunting-ground, but the numbers now are greatly reduced. The last accessible statistics put the total number of skins at about four thousand a year. The natives, who hunt on Saanach Island, go there every winter, and their sufferings must be terrible. They cannot make fires even to cook their food, because fires

FIG. 518.— Sea-otter (*Enhydris lutris*).

scare away otters, and here they live for weeks, the thermometer below zero, in a northerly gale of wind. The otters have the senses of hearing and smell acutely developed. A fire five miles away alarms them, and even the tracks of man upon the beach must be washed by many tides before they will land on that part. They are taken by surf-shooting, by spearing-surrounds, by clubbing, and by nets. Surf-shooting is accomplished with rifles. As soon as a head appears, even a thousand yards out, it is shot, and then the surf brings in the body. Spearing-surrounds are more difficult. A dozen or twenty boats start out in a long line, and when one sees an otter, he rows quickly to it. It usually dives before he can spear it. He stops right where it was last seen, and the others scatter round in a circle, each on the lookout. When the otter appears again, the same tactics are repeated, until it is at last thoroughly exhausted. Clubbing and netting need not be described.

There are three or four species of true otters, but only our own species distributed over nearly the whole of our country, but nowhere abundantly. Still, immense numbers are taken, the Hudson Bay Company selling over eleven thousand in London in 1873. They frequent streams and ponds, and are caught with steel traps. They are busy and voracious animals, but they have their times of relaxation, and then coasting seems to be their highest enjoyment. Sometimes this is done on wet, clayey banks in summer, or on the snow in winter. Says Godman: " Their favorite sport is sliding, and for this purpose in winter the highest ridge of snow is selected, to the top of which the otters scramble, where, lying on the belly with the fore feet bent backwards, they give themselves an impulse with their hind legs, and swiftly glide head-foremost down the declivity, sometimes for a distance of twenty yards. This sport they continue apparently with the keenest enjoyment, until fatigue or hunger induces them to desist."

Our American badgers are found in the west, and owing to their mode of life their habits are but little known. They live underground either in burrows which they dig for themselves, or in those of the prairie-dog which they readily enlarge for their needs. They are very secretive in habits and live almost as much below the surface as does a gopher. In some places they are very abundant, and their holes dot the ground so that it is almost impossible for one to ride over it. Prairie-dogs are their especial food, but they will eat flesh of almost any kind. They will try to avoid danger, but when brought to bay they are filled with courage supported by great strength and a formidable armature of teeth and claws. They possess great vitality and will prove awkward customers for dogs of the largest size. The European species is equally plucky, and the sport of badger-baiting with dogs has given us the verb, to badger.

Our badger and the European species are strong-smelling, but they are far excelled in this respect by the stinking badger of the East Indies. Mr. Forbes, speaking of this species, says: " Another slow prowler very often made my evening hours quite unbearable by the intensely offensive odor with which, even in its most inoffensive frame of mind, it hedged its crepuscular walks for at least a mile around. It was no use to frighten it away; for if its equanimity were disturbed, it did not haste to its lair as one could have wished. It thickened, instead, the very air with a malignant scent that clung to one's garments, furniture, and food for weeks. . . . The native has a superstition that if a man have fortitude enough to eat its flesh, he will have become proof against sickness of every kind."

This matter of smell brings us at once to the skunks, a group confined solely to the New World, and with which pages could easily be occupied; but since we have not the space, we will omit any description of the nature of the most prominent characteristic of these animals and merely say that the odor is destroyed by ammonia. There is, however, one side of the skunk that demands a moment's attention. In 1874 the question was

Fig. 519. — Wolverine, or Glutton (*Gulo luscus*).

raised, does the bite of the skunk cause hydrophobia? and many cases were cited where this disease and death had resulted from the bite of this animal. It is, however, certain that not invariably is the bite of a skunk productive of rabies or any other serious affection, while on the other hand it is equally certain that such cases have occurred. To explain this two hypotheses have been advanced; one that the skunk communicating the

virus had been bitten by a rabid dog, or that possibly some derangement
of the odoriferous system had occasioned the disease.

The wolverine, or glutton, well deserves the latter name, as well as its
Latin equivalent, *gulo*, even when we dismiss some of the extravagant
tales of the older writers on natural, or better on *unnatural*, history. Says
Dr. Coues: "Probably no youth's early conceptions of the glutton were
uncolored by romance; the general picture impressed upon the susceptible
mind of that period being that of a ravenous monster of insatiate voracity, matchless strength, and supernatural cunning, a terror to all beasts,
the bloodthirsty master of the forest. We cannot wonder at the quality
of the stream, when we turn to the fountain-head [Olaus Magnus, A.D.

Fig. 520. — Weasel (*Putorius vulgaris*), above; ermine (*Putorius ermineus*), below: both in summer dress.

1562] of such gross exaggeration. We find it gravely stated that this
brute will feast upon the carcass of some large animal until its belly is
swollen as tight as a drum, and then get rid of its burden by squeezing
itself between two trees, in order that it may glut itself anew." In
reality, the wolverine is a strong animal, with an enormous appetite, and
besides, a spirit of wantonness scarcely exceeded in the whole animal kingdom. It destroys more than it can eat, and then defiles the rest so that
scarcely a dog will eat it. It enjoys hiding things, and cases are reported
where they have plundered a camp or a settler's cabin and carried off and
hidden every movable thing — blankets, kettles, cans, knives, guns, etc. It
follows the trapper on his rounds, pulls the log-traps to pieces, and steals
the bait from behind. Their cunning makes it a difficult task to capture

them, and yet they must be captured before one can trap successfully in their region; for they will tear the traps to pieces as fast as the trapper can build them.

The wolverine is but a sub-plantigrade; the remaining carnivores, even those of the bear series, walk upon the tip of the toes. Of these forms the weasel and the stoat, or ermine, are well known — small, long-bodied forms, common to both Europe and America, and which turn nearly or completely white in winter in the more northern portions of their range. They are both savage, bloodthirsty animals, with a peculiar 'snaky' appearance, which renders them far from attractive. The stoat is the worst of the

Fig. 521. — Mink (*Putorius lutreola*).

two from this point of view. It is also the stronger and tougher. It destroys for the simple pleasure of taking life. In the hen-coop it will destroy every inmate in a night; in a rabbit-warren it follows one rabbit after another, killing many, but never leaving the pursuit of one to take up that of another until the first has been pierced with those sharp teeth. Still, this very habit of destructiveness makes both the weasel and the ermine the farmer's friends. When they get into a barn, they kill off the rats and mice, while in the field they make sad havoc with the field-mice and moles.

The ferret is a native of Africa which has been introduced as a ratter into almost all parts of the world. It is now scarcely known outside of domestication; but no matter how tame it may be, it is always a stupid animal, except when excited by game. It is a very near relative of the polecat (the name comes from Polish cat), which has the doubtful honor of being the worst-smelling animal in all Europe. It is also known as fitch and foul-mart, and furnishes the well-known glossy black fitch fur of commerce. It is as insatiate as the rest.

Our mink stands next, — the skunk of our animals in the strength of the smell, but it is not so insufferably perfumed as is that little black and white 'enfant du diable'; and besides, its aquatic habits render its effluvium less apparent. Minks occur in the neighborhood of water all over North America. They make their burrows in the banks of the streams and ponds, in which they seek their food of fish, molluscs, cray-fish, reptiles, and frogs:

FIG. 522. — Asiatic sable (*Mustela zibellina*).

but they are perfectly willing to vary this diet with mice and moles, a possible musk-rat, and an occasional visit to the neighboring poultry-yard.

The American sable, or pine-martin, is one of our most valuable fur-bearing animals. It occurs from the northern part of the United States north to the limit of trees. It differs from the forms already enumerated, in that it shuns civilization, and but rarely, if ever, does it visit the chicken-coop. A remarkable feature is its periodicity. Once in every few years they come out in great multitudes, as if their retreats were overstocked, and in the early years of this century the annual catch of this animal amounted to one hundred thousand. It is chiefly trapped in the winter months, when the skin is in the best condition, the trap being made of brush and stakes, with a log for a dead-fall.

The European birch-martin and pine-martin may be dismissed with mere mention, and their relative, the Asiatic sable, the most valuable of all the fur animals, needs but little more space. It is an Arctic form, found in all the more northern parts of the Old World, but most abundantly in northeastern Siberia. It is most valuable in its winter pelage, and the sufferings the trappers go through, in the rigors of a Siberian winter, can better be imagined than described. Skins of Asiatic sable vary much in quality. Inferior skins may be purchased for about twenty-five dollars, but those of the finest quality will bring two hundred dollars, or even more.

The last of these weasel-like forms is blessed with a multiplicity of common names, — pennant, pekan, fisher, black cat, black fox, marten, and the like. Black fox and black cat may at once be thrown aside, for the animal is neither fox nor cat. Neither is it a fisher; for all accounts agree that while it is fond of fish, it does not catch these animals. It is a marten; but that term is shared by several other animals. Pekan is probably of Indian origin, and is the name most used. The pekan is a strong animal, and feeds on small quadrupeds and birds. The porcupine with his spiny armor is no match for this animal, and all the hunters agree in saying that this animal is an especial delicacy. Dr. Merriam quotes an instance where the intestines of one of these animals "contained hundreds of porcupine quills, arranged in clusters, like so many packages of needles, throughout its length. In no case had a single quill penetrated the mucous lining of the intestine, but they were apparently passing along its interior as smoothly and as surely as if within a tube of glass or metal." The strength and cunning of the pekan makes it almost as great a nuisance to the trapper as the wolverine. It tears the traps to pieces in the same manner. They do not, however, seem to show that malicious spirit that is so characteristic of the glutton.

The next great series of carnivores is that of which the dogs may be regarded as the typical members; and of these the long roll of foxes first comes up for mention. A fox may be known by its erect ears, sharp muzzle, bushy tail, and elliptical eyes much like those of a cat. In our eastern country but two species occur, and of these but one is at all common. The gray fox does not appear to thrive in the vicinity of man, and it has now become scarce in regions where once it was very abundant. Much more common is the cunning red fox, which shows the crafty habits of the well-known Reynard of the Old World. The red fox is a pretty animal, with an intelligent countenance, which would make the animal many friends were not its habits so bad. If when tamed it would only lose its treacherousness, and if when wild it would let the hens and lambs

alone, it would stand in far better favor. As it is, it is hunted wherever
it occurs. In the chase it is full of wiles. It will double on its track,
take to the water, and do every conceivable thing to throw the dogs off the
scent, and at last it seeks its burrow, which may be placed among the
rocks, where it is next to impossible to dig the animal out. Some relish a
fox-hunt — not one of those imitations which have been introduced at New-
port and on Long Island, where a captive fox is let loose, but a real hunt
where a wild fox is the object of pursuit. I have engaged in one, and have
had a surfeit. Their track is found in the light snow, and the dogs are
set loose. On they go, their noses to the ground, while the human hunters

Fig. 523. — Gray fox (*Vulpes cinereo-argentatus*).

take themselves to some favorite runway of the animal, and there await
his appearance. Sometimes the fox makes his journey a short one; but
more often he prolongs it for miles, and the chase becomes a question of
endurance between dog and fox, and a severe trial upon the muscles of
the men. After all, the fox may escape; but even if caught by dogs or
gun, there is little satisfaction, except to the farmer whose poultry-yard
has been plundered, in gazing on the last struggles of the sagacious little
beast.

Among the other foxes are the kit, or burrowing-fox of our western
territories; the common fox of Europe, whose cunning and duplicity are
scarce exaggerated in the mediæval tale of Reineke Fuchs, and which
furnishes sport for the British squire and yeoman; and the Arctic fox

of all the countries around the north pole, an animal which like many other boreal forms, turns white in winter.

The differences between dogs and wolves are but slight; both are placed in the same genus, and it is more a question of nomenclature than of structure, as to whether any species be regarded as a dog or a wolf. Indeed, it is highly probable that there is much wolf blood in some of the breeds of domestic dogs, together with that of jackals and foxes. The true wolf, found both in Europe and America, is the most formidable of the lot,

Fig. 524. — Wolf (*Canis lupus*).

and the only form that is dangerous to man. It is now exterminated in most of western Europe; but vast packs still remain in the forests of Russia. A single wolf is a cowardly creature; but in a pack they are the personification of ferocity, and tales abound of people being chased by these animals. Wolves have from time immemorial been the subject of superstition. The belief in were-wolves — persons capable of turning themselves into wolves at pleasure, and satisfying their appetites with human flesh — are well known, and in the same line is the following: —

" In Arcadia there was a certain race and house of the Antæi, out of which one ever more must of necessitie be transformed into a wolfe; and when they of that familie have cast lots who it shall be, they use to accompanie the partie upon whom the lot is falne, to a certaine meere or poole in that countrey: when he is thither come, they teene him naked out of all his clothes, which they hang upon an oke thereby: then he swimmeth over the said lake to the other-side, and being entered into the wildernesse, is presently transfigured and turned into a wolfe." Here he dwells, keeping company with the other wolves for nine years, after which, if he have totally abstained from human flesh, he may swim back across the lake, and resume his former shape and condition, save that he is nine years older.

FIG. 525. — Coyote, or prairie-wolf (*Canis latrans*).

The wolves themselves were also regarded with equal dread; they were endowed with supernatural powers, and "it is commonly thought likewise in Italie, that the eyesight of wolves is hurtfull, in so much as if they see a man before he espie him, they cause him to lose his voice for the time."

The coyote, or prairie-wolf, is a smaller and more timid animal than the true wolf, but it has the same habit of hunting in packs. Night is the time for its greatest activity, and then the chorus of howls and barks drowns every other noise. The coyote is a cowardly but cunning beast, ranging from the plains to the Pacific, and a familiar element of every picture of life in all that region.

Besides the marsupials already mentioned there was but one terrestrial mammal in the great Australian continent at the time of its discovery by man. This was the dingo, a dog which ran wild through all the island, leading a predaceous life, and which was occasionally reduced to a state of semi-domestication by the natives of the country. Whether it was indigenous, or was introduced from some other region, is a problem as yet unsolved, although if the latter supposition be the true one, the date of its introduction must have been exceedingly remote, for the remains of the dingo are found fossil in strata of quaternary age.

Of the domestic dog we have but little to say. The subject really demands a volume, but our space will only admit a reference to the most peculiar form, — the Japanese pug. Long domestication has resulted in great degradation of structure; and an old dog, instead of the forty-two teeth typical of other dogs, may have these reduced to sixteen. Were the dog found wild, no naturalist would hesitate to place it in a distinct genus, if not in a distinct family, from the other dogs.

Last of the terrestrial carnivores comes the group of cats, civets, and hyenas, the latter of which are the most dog-like, and hence need consideration first. Hyenas are far from pleasing animals. They are very strong, and furnished with teeth adapted to crush the hardest bones. They will break the shin-bone of an ox with their jaws, while they are able to carry off a weight of seventy or a hundred pounds. They are as voracious as is possible, and will gorge themselves whenever opportunity offers, rivalling the glutton in this respect. Yet, strong as they are, they are cowardly, and take night for their prowlings; but if brought to bay, they make formidable antagonists. Hyenas play a good part in the hotter portions of Africa and western Asia; for they are scavengers, and they rid the country of the carcasses of game, which would otherwise breed pestilence. But this habit has one drawback; for they will desecrate cemeteries, digging the corpses from a depth of five or six feet. In the day-time they live in holes in the rocks, caverns, and the like. It is hardly necessary to refute the statement that the hyenas are untamable, for it has often been shown to be false.

The civet cats are mostly Asiatic, one species occurring in eastern Africa. All are remarkable from their secretion of odorous substances, and civet is one of the most highly prized perfumes of the east, though so rank as to be disagreeable to the senses of Europeans. Civet is a secretion from certain glands in the genital and anal region, which collects in a pouch, and serves to call the sexes together at the pairing season. Excitement of any sort increases the amount, and so, to obtain it from the civet of Africa and the rasse of Japan, the animal is confined in a small cage, without chance of turning round, while the pouch is scraped with a spoon. This enrages the animal, and an increased amount is secreted, but there is no chance for the angry beast to use its teeth. In the case of the genette,

Fig. 526. — Brown hyena (*Hyæna brunnea*).

which has no pouch, the musk is scraped from the bars of the cage, on which the animal rubs itself. Civets and genettes are too irritable to make very desirable pets, and yet they are frequently kept about the house, like cats, to free it from rats and mice.

In these respects they are excelled by many of the ichneumons, animals almost as characteristic of Africa as the civets are of the Asiatic fauna. The ichneumon of northern Africa and Spain is one of the best-known, and its habit of destroying vermin of all sorts brought it into high repute with the ancient Egyptians, who embalmed its body. In India the Indian ichneumon, or mongoose, takes its place, and its value as a destroyer of rats, snakes, and other pests makes it a favorite. It does not hesitate to attack the most venomous forms, and some say that it knows some specific against their poison, and that when bitten it rushes off to the woods and eats the plant which is to counteract the virus. This, however, seems to be an erroneous idea like that of the Indian specific against the bite of the rattlesnake. The mongoose probably owes its immunity solely to its audacity and agility. It avoids their blows with

Fig. 527.—Indian ichneumon, or mongoose (*Herpestes griscus*).

great adroitness, and tries to plant its teeth in the head of the cobra itself. All of the ichneumons make interesting pets, as they are intelligent and active.

Near relatives are the suricates of Africa, which are even more interesting in confinement. Frank Buckland had a tame specimen of whose habits he has given an interesting account. It was full of curiosity, and examined everything. At breakfast time it would make its appearance at the table, and climb upon it if not fed in due season. It made a most amusing sight when it would " smell of the eggs, and swear at them in its own language." After breakfast there was usually a fight with the monkeys; for the suricate would try to steal their food. He had found

out that the cat's claws were sharp, and so when robbing tabby of her
dinner, he adopted different tactics from those used when stealing from
the monkeys. The latter he fought face to face; but when after the cat's
dinner, he ruffed up his fur, and pushed himself backwards up to the cat,
who, much disturbed at this, would give him a dab with her claws. Alas
for her! the blow always cost her her dinner; for the suricate, reaching
his nose through under his body and between his hind legs, seized the
morsel, while the cat's attention was directed to the blow which fell comparatively harmless on the ruffled fur. This animal, like its relatives, was
inquisitiveness itself. Every hole it could find it insinuated itself into,
and its great delight was to get into a shoe or boot, and then to push it
along over the floor, as if it were impelled by some invisible machinery.
It rarely walked, but went about the house on a gallop or canter.

The highest and the most graceful of all the carnivores are the cats, a
family which embraces not only our familiar puss, but the lion and tiger
and a host of smaller forms as well. All are cat-like; they have the same
silent tread, the sharp claws, when not in use, being drawn back into
velvety pockets, where the ground will not injure the points; all are
graceful in every motion, and our domesticated kittens are no more agile
and playful than are those of the tiger or lion; all are bloodthirsty and
treacherous. Fifty odd species are known, thirty-five of which are from
Asia, while America has about a dozen.

If size and prowess are criteria, the tiger stands an easy first among all
the cats. In both respects it stands ahead of the lion, which for centuries
has been termed the 'king of beasts.' Its home is India, but it ranges
far outside that country; in India, however, it is best known, and there it
furnishes the highest class of 'sport' for the English officers. Tiger-hunts
are relieved from the curse of many other so-called sports; for every tiger
killed is a benefit to the country; but tiger-hunts are nearly as dangerous to the pursuers as to the pursued; for it not infrequently occurs that
several fall victims to the beast before the rifle lays him low. Tigers
delight in the thick jungle, whence they issue at intervals to carry off the
cattle of the villagers, or, perchance, a human being. There are several
ways of hunting the tiger: one is to tether a goat or kid in some place
which they frequent, and then await his nocturnal approach, and shoot
him from a platform in the trees. This, however, is tame sport beside
that of driving him from his lair in the jungle, and then hunting him from
horseback, or from a howdah on the elephant.

In killing an ox or a buffalo the tiger usually tries to seize it with its
jaws by the nape of the neck and dislocate the backbone by a wrench
while the body is held fast by the paws. It has, however, been known to

ROYAL BENGAL TIGER (*Felis tigris*).

strike a buffalo on the back with the paw with sufficient force to bring it to the ground.

The lion ranges from western Hindustan to the southern extremity of Africa, and is a form too familiar to take more of our space. There is but a single species, but this acquires different characters in different parts of its range.

FIG. 528. — African lion (*Felis leo*).

Of a smaller size is the leopard — spotted lion the name means — of Asia, one of the most beautiful of animals. It frequents the forests and can climb like a cat. As a rule it is not disposed to touch man, but after it has once tasted human blood it becomes a confirmed man-eater, and indulges its appetite as often as opportunity offers. It is a strange fact that the peculiar odor of small-pox patients is especially attractive to it. It is usually caught with pit-falls or with spring-cages baited with kid.

About the same size is the cheetah, or hunting-leopard of India, the only one of the large leopards that is of service to man. It is tamed without much difficulty, and is used in the hunt, being taken to the field with a hood upon its head. An extract from the pages of Gordon Cumming will show the way in which it is used. The party had discovered a herd of deer. "A cheetah was now unhooded, and on seeing the deer, he at once glided from the cart, and taking advantage of every tuft of grass and inequality in the ground, he crept towards his prey. The deer were

Fig. 529. — Leopard (*Felis pardus*).

meanwhile lazily watching us as we went on without halting, and the poor beasts were only aware of their danger when the leopard made his rush. There was a wild scurry, but the cheetah was among them, and as the herd cleared off, we saw him lying with his teeth in the throat of a goodly buck. His keeper now came up with a wooden ladle and a knife, and cutting the deer's throat, he caught the blood in a spoon, into which in a few minutes the cheetah thrust his nose; and while he was lapping the blood, the hood was slipped over his eyes, and he was secured and replaced in the cart."

The other of the Old-World cats must be summarily dismissed; the ounce, the various tiger-cats, the wild-cat, the jungle-cat, and all the rest, can only be mentioned. Even domestic pussy can have but a few lines. It is uncertain from what wild species this form has descended. It was tamed as long ago as any records extend, and it may be that the wild-cat of the Old World was the ancestral form, though the probabilities are against this view. It is a remarkable fact that the cats vary less in domestication than does the dog. Of the latter animal the breeds are numerous; but with the cat there are but few distinct types, and some of these will not breed true. Cats occasionally take to the forests, forget all the long line of domesticated ancestors, and live a life as wild as any other form.

Of the cats of America, the puma is the most widely distributed. It ranges from Patagonia north to Hudson Bay. It has a list of names almost as large as its range; cougar, panther, painter, catamount, carcajou, American lion, and a dozen others are applied to it. It figures largely in all the romance of our American forests, and as a result no little exaggeration has been gathered about it. Dr. Merriam has some interesting notes on these points. As to size, after giving some fabulous accounts, he says: "To those that are inclined to credit such statements I have only to say: measure off eleven feet on your floor; place the largest panther you ever saw on this line, and then tell me on what part of the beast you would 'annex' or 'splice on' the three or more additional feet." He is equally sceptical as to the celebrated scream of the panther; its jumping from its lurking-place in the branches of a tree to the back of a passing deer; and the oft-repeated stories of them carrying off their prey slung upon the back. He also denies the ferocity of the beast, and truly adds that "it is now well known that the panther is one of the most cowardly of beasts, never attacking man unless wounded and cornered; that it is unnecessary to do more than contradict the popular impression to the contrary."

In reality the puma steals upon its prey (largely deer) as closely as possible, and then when thirty or forty feet away, begins its leaps. Usually there are one or two preliminary leaps before the final one, which is sometimes of almost incredible length. Twenty feet is not an uncommon spring on level ground, and leaps of nearly forty feet have been measured. If the panther does not strike its game the first time, it but rarely makes an attempt to follow it.

A larger animal is the jaguar; indeed, it stands second only to the lion and tiger in size and strength. Fortunately for us it does not range farther north than Texas and Louisiana, but south of those states its distribution is almost co-extensive with that of the puma. It is a fierce animal, and its great strength makes it the terror of all tropical portions of our continent, and yet a drove of peccaries will master one, if no trees are near, tearing it to pieces in their savage way. In its habits the jaguar differs but little from the other large cats. It feeds on whatever of fortune may come in its way, — deer, monkeys, peccaries, squirrels, and the huge and clumsy copybara.

The lynxes are smaller cats belonging to the northern hemisphere, and the species figured is common to both continents. From their smaller size they are compelled to take to smaller game than the panther; and squirrels, grouse, porcupines, and other inhabitants of the forest are their prey.

Fig. 530. — Lynx (*Felis borealis*).

They will not willingly attack man, but when cornered, their sharp claws and teeth make them antagonists not to be despised. The common wild-cat or bay-lynx is but a variety of the form figured.

The seals and their allies are the most aberrant of all the carnivores, following out as they do a line of which we saw traces in the otter and sea-otter. In the present group the whole body is fitted for life in the water; the feet are transformed into flippers, admirably adapted for natatory locomotion, but of little use on land; and in the seals themselves the hinder pair of limbs are turned backward, so that they and the small tail form a pretty perfect substitute for the flukes of a whale. Indeed, by some these latter organs are supposed to have had their origin from members like the flippers of a seal, a view which though plausible is not very probable.

Of the present group, the huge walruses of the Arctic seas are the most terrestrial; for, by doubling their hind flippers under them, they are able to make a slow progress upon ice-floes or on the ground. The most striking features about them, however, are the long tusks which project downwards from the upper jaws. These are said to be used in digging up the mud in the search for the shell-fish on which these animals feed, and for climbing on the ice, as well as for organs of offence and defence. The walrus goes in large herds, a feature in its habits which makes it a profitable object for pursuit; for though the oil tried from its blubber is less valuable than that from the seals, still so much may be obtained from a herd of these animals as to make up the difference. Besides, the tushes are valuable, and especially so since the supply of elephant ivory has so diminished. A full-grown walrus may weigh considerably more than a ton.

Of the seals there are two groups, one difference between which is indicated by the names, eared seals and true seals, the latter having no external ear. To the eared seals belong the sea-lions and the fur-seals. They occur in the Antarctic seas and in the north Pacific, but none are found on the northern Atlantic shores. The large northern sea-lion may reach a length of thirteen or fourteen feet, and a weight of a thousand pounds. It is a lion by courtesy. It can utter a loud roar, but in few other respects, certainly not in bravery, does it resemble its prototype. Indeed, before man, it is a most timorous animal, and the Aleuts have no difficulty in frightening these animals and driving them to the slaughtering places with flags and umbrellas. To these people they are of the greatest value; for like the cocoanut-tree of the tropics, they furnish almost every necessity of life, — meat, oil for light and heat, sinews for thread, leather for clothing and shoes, etc. Even the digestive tract is utilized, the stomach serving for reservoirs for liquor; while the intestines furnish material for water-proof clothing, as well as a substitute for glass in the windows.

To us the fur-seals are more important, especially since sealskin became so fashionable. Most important is the northern fur-seal, whose headquarters is at the celebrated Fur-seal or Prybilov Islands, which are estimated to have a seal-population of nearly five million. A while ago there was an indiscriminate slaughter, but in recent years the United States government has leased the islands to the fur companies, who are compelled to keep the slaughter within such limits that the annual increase will make good the loss. In the early summer the seals begin to arrive, the males coming first. They march up from the water, their heads erect, and raised about three feet from the ground, and each selects a spot for breeding, which he must defend against all comers. This affair settled, the females, about a month later, begin to arrive, and then ensues a scene well described by Elliott. As the cows draw near, the bulls go to the water and coax the females to their seraglios. As soon as one is enticed, the male starts for the water to secure another, but as soon as his back is turned, a neighbor has taken the first wife off his hands, and then a battle ensues, which not infrequently results in the poor little female, not a

Fig. 331. — Harp-seal, or Greenland seal (*Phoca vitulina*).

quarter the size of her lord, getting badly used. They stay on shore but a short time, — the males until August, the females a month or two later; for they have to care for the cubs, which are born soon after the mothers appear. The young males furnish the best skins, but before these are ready for the market they have to pass through many processes to give them that soft appearance so much admired.

There are fur-seals in the Antarctic seas, but their fur is not so valuable as that of the Alaskan species; and besides, their capture has not been regulated, so that now, where a quarter of a century ago they were abundant, they have almost been exterminated, and to-day a sealing-voyage does not have the profit that it formerly did.

Of the true seals there are many species, but none is more familiar than the little harbor-seal of our eastern coast, which occasionally makes its way up the St. Lawrence to Lakes Ontario and Champlain, and even to Onondaga Lake, near Syracuse. There are few sights more pleasing than a number of these seals at play on some rocky shore. In their gambols they seem to exhibit a fondness for sport and humor which one would not expect in such animals, but their mournful howls at night display another side to their character. The fisherman, too, looks at another aspect of these animals; for they are expert fishermen themselves, and, besides, they do not hesitate to rob the nets of everything they may contain.

Farther north there are other seals, and all are as important to the Eskimo as is the sea-lion to his Aleut cousin. A little farther south, at the mouth of the St. Lawrence, is the great Atlantic seal-fishery, and it is estimated that there nearly a million of these animals are slaughtered annually for the sake of the oil and skins: their fur is of but very slight value. All the seals are intelligent, and, as every zoological garden bears witness, they may be taught a variety of tricks.

Last of the seals are the hooded-seals and the sea-elephants, almost all the existing pictures of which are faulty in regard to the peculiar structures of the head, from which they derive their names. The hooded-seal, which lives in the north Atlantic, is usually said to have a hood, or cap, on top of the head. In reality, this is an inflatable proboscis, overhanging the mouth, and is found in the males only. What its function is has not yet been determined. In the sea-elephants this same proboscis occurs likewise in the males, but is much more fully developed. At a state of rest it may be fifteen inches long, but when excited it elongates to a considerably greater length. This proboscis, recalling that of the elephant, gives the animal its name. There are two sea-elephants, one occurring on the Pacific coast of North America, the other inhabiting the seas around the South Pole. They are very large animals, the males averaging fourteen feet in length, and occasionally measuring twenty, from the tip of the proboscis to the end of the toes.

PRIMATES.

In all the older works on natural history the lemurs, monkeys, apes, and man came first, and then, in descending scale, followed the immense host of animals which have been so summarily enumerated in the foregoing pages. Hence the name Primates (*primus*, first) was very applicable to them. Now the old system has been turned end for end: we ascend rather than descend: and the result is that the first are last, and the last first. Still, we retain the old name, for no one would think of changing it to ultimates. Linné gave the name Primates, and included in it the bats, as well as the forms enumerated above, and later, Cuvier divided it still farther, and placed man in a group by himself. Every naturalist to-day regards this as a false step. In anatomical characters man differs but very slightly from the apes, even if there is a great intellectual gap between the two. Psychological characters, however, cannot be taken into consideration in classification, and man must remain the leader, but still a member of the same group with the apes and monkeys.

Fig. 332. — Slender loris (*Stenops gracilis*).

Except to the naturalist, the lemurs possess but little interest in comparison with the other members of the group of Primates. The species all belong to the Old World, and they seem to have remained at a low intellectual stage. The true lemurs are confined to Madagascar and the Seychelles and Mascarenes. A few feed on birds, of which they suck the blood only, but vegetable matter seems to be the mainstay of all. Some like grass, while others plunder fruit-trees, and are especially fond of dates and plantains. Many of the species are regarded by the Malagassy with a wonderful amount of veneration and awe.

Though occurring in Asia and Africa, the slow lemurs are wanting in the Malagassy fauna. They are rightly named: for they are the incarnation of slowness: even a sloth is quicker than they. Take, for instance, the species represented in the cut, the slender loris, or Ceylon sloth. It is sluggish and inactive even at night, when it hunts for its prey. It eats rice, fruit, and vegetables; but it is also fond of flesh and insects, and in capturing its animal food its mode is characteristic. It approaches in a stealthy manner, and then extends one of those long arms so slowly that the prey is not alarmed until it is seized by that peculiar hand. In sleeping it assumes the most uncouth and grotesque postures. The eyes are very peculiar, large and with a very small pupil. Indeed, the Cinghalese

Fig. 533. — Spectre (*Tarsius spectrum*).

recognize them as something strange and wonderful; and when a lover wishes a charm to render him successful in affairs of the heart, he catches one of these animals, holds it before the fire until the eyeballs burst, and then takes these organs, which are supposed to be of great efficacy so prepared.

The head of a cat, the teeth and tail of a squirrel, the hands of a miser, and the feet of a monkey have been enumerated as characteristics of the strangest inhabitant of Madagascar, the aye-aye, so called from the noise which it makes, and not from any Malagassy exclamation of surprise, as is often stated. It is a strange creature which was formerly regarded as a rodent, but is now assigned a place near the lemurs. Strangest of all are its long, skinny fingers, the middle one long and scarcely more than skin and bones. This finger is a most important instrument to the animal. In drinking and in eating juicy fruits it plays the part of a spoon, or, better, of the Chinese chop-sticks. It also serves as a probe and a fork in extracting worms from the burrows in the wood. The aye-aye is a nocturnal animal, and this fact partly accounts for its rarity in collections. This rarity is, however, to be attributed also to its limited distribution on Madagascar, and also to the veneration in which it is held by the natives.

Figure 533 represents the strange spectre or spectral lemur of the Malay Archipelago, in which the eyes of the slow loris are even exaggerated. It is a small animal, the body being only some six or seven inches in length. It lives in the dense forests, making its nest in the hollow roots of the large bamboos, and climbing about with as much agility as any tree-frog. In its arboreal evolutions it derives great assistance from the large suckers on the tips of the fingers and toes, which strongly recall those of the animals just mentioned. One has but to compare this figure with Figure 336 to be struck by the similarity. This animal is nocturnal, and for this life the eyes are especially adapted. It feeds largely upon insects.

The other series of Primates leads in a nearly straight line from the marmosets to man, while the lemurs just noticed are an offshoot from the main stem in the direction of the spectre and aye-aye. Of the present line the marmosets — inhabitants exclusively of South America — are unquestionably the lowest. In both habits and appearance they are much like the squirrels, but are even smaller than these, being the most diminutive of all the monkey tribe. Their thumbs are not opposable to the other fingers, and the finger-tips are armed with claws rather than with nails. The body is covered with long, soft hair which frequently upon the head is developed into curious ornaments, as shown at the left of our figure of these animals, or into a long mane, recalling that of the male lion. The body terminates with a tail nearly twice as long as itself; but this member is not as serviceable as it is in some of the other New-World monkeys, for it cannot be coiled around objects.

In captivity they make very interesting pets, but more from their activity and vivacity than from any intellectual superiority. They are timid, but come to display considerable affection for those who care for them. Their curiosity is easily excited, and then they are at their best: for the head is then in active motion, peering this way and that and trying to solve the mystery. An oft-told but interesting story is that related of one kept by the French naturalist Audouin. It exhibited great fear when a wasp was presented to it, though it enjoyed nothing better than to catch flies and other insects. This led to the experiment of showing it the colored plates in a work on natural history. At the pictures of a cat and a wasp it recognized its enemies, and became greatly terrified, while it quickly

Fig. 534. — Marmosets, or sagouins (*Hepale*).

precipitated itself upon the figure of a beetle or a grasshopper, as if to seize the objects represented there. Marmosets are frequently brought to our northern countries, but they do not thrive over-well. The weather is too cold, and like the rest of their tribe, they show a great tendency towards consumption, which affects them exactly as it does man, and terminates with the same sad results. In their native country they live on insects, birds, eggs, and fruit. Some inhabit the virgin forests, while others prefer the plains in the neighborhood of the habitations of man. The Indians shoot them with poisoned arrows, and afterwards restore them with salt, or catch them in cages and traps. Besides being kept as pets by the Indians, they are, together with other monkeys, used as food. This may at first seem a depraved taste, and we cannot forbear a short extract from the pages of Hæckel, which, though written of another form, is equally applicable to all. "Among the dainties I have mentioned above as the result of my sport, I spoke of monkeys," says the great Jena naturalist. "I found this noble game excellent eating, either fresh and baked, or pickled in vinegar; and I began to suspect that cannibalism was, in fact, a refined form of *gourmandise.*"

Besides the marmosets — they are also called uistitis and sagouins — South America nourishes a long series of monkeys, — the titis, couxios, uakaris, — which are of too little interest to occupy more space than is necessary to say that their names are of Indian origin. The howlers are more important, for they form a part of every picture of tropical America. Every traveler up the Amazon or the Orinoco devotes pages to them, and tells how every night and morning the forest resounds with their loud voices, and how the whole chorus has a leader, who directs their vocal performances. They can be heard for miles; and an examination of their throat structure reveals a peculiar apparatus for producing a noise. Besides well-developed vocal cords, the windpipe has large sacs — sounding-boards in function — which tend to strengthen the sound, so that the loudest noise is made with comparatively little effort. They move in large bodies through the woods, passing from tree to tree in a regular follow-my-leader style, and in these journeys the tail is almost as useful as the hands or feet; for it can be coiled around a branch tightly enough to support the weight of the animal, and not infrequently the hold will be retained long after the animal has died from the effects of a gunshot. In confinement they are not interesting; for they never become tame, and always retain a surly disposition, no matter how kindly they may be treated. Some of the howlers are the largest of the New-World monkeys, the black howler having the body twenty-eight inches long.

The sapajous have the same prehensile tail as the howlers, but they lack the complex vocal apparatus. They are far more interesting in confinement than the others; and as they bear constraint and a cold climate well, they are among the most common inhabitants of menageries and zoological gardens. They are very intelligent, the capuchin leading all the rest in this respect, and showing a degree of understanding almost human. In captivity this species is constantly engaged in investigation, and besides, it shows other human traits. Ridicule is as dreadful to it as to any schoolboy.

The spider-monkeys — so called from the extreme length of their legs — are first-cousins to the sapajous. In them the prehensile capacities of the extremely long tail reach their highest development. In their passage through the trees this member is whipped around some branch to support

Fig. 535. — Sapajous (*Cebus*).

the body, or even extended some distance ahead to bring some bough of an adjacent tree within reach of the paws; or again, the same fifth hand may grasp the fruit on which the animal is feeding. The spider-monkeys are distributed over the whole Amazon basin, some of the members extending north into the southern Mexican states. The different species have their vernacular names, among which the most common are miriki and coaitá, the latter not to be confounded with the coaiti mentioned among the beasts of prey. Bates, who lived many years on the Amazon, gives the following note on one species : —

"I once saw a most ridiculously tame coaitá. It was an old female, which accompanied its owner, a trader on the river, in all its voyages. By way of giving me a specimen of its intelligence and feeling, its master set to and rated it soundly, calling it scamp, heathen, thief, and so forth, all through the copious Portuguese vocabulary of vituperation. The poor monkey, quietly seated on the ground, seemed to be in sore trouble at this display of anger. It began by looking earnestly at him, then it whined, and lastly rocked its body to and fro with emotion, crying piteously, and passing its long, gaunt arms continually over its forehead ; for this was its habit when excited, and the front of its head was worn quite bald in consequence. At length the master altered his tone. 'It's all a lie, my old woman ; you're an angel, a flower, a good, affectionate old creature,' and so forth. Immediately the poor monkey ceased its wailing, and soon after came over to where the man sat."

An allusion must be made to the oft-described way in which the spider-monkeys cross a stream. If it be narrow, they leap across it, trusting to luck and their tails to find a landing-place on the other side ; if it be wider, they form a living bridge by joining themselves into a long chain, and swinging from a tree until the end monkey can reach the branches on the opposite side. Over this bridge the whole troop travel, then the bridge itself is broken at the starting-point, and then all are safe on the farther side. All of the spider-monkeys are eaten, and their flesh is highly praised.

The remaining monkeys and apes all belong to the Old World. They stand on a higher plane than those of America so far as approximation in structure to man is concerned. They have the same number of teeth, the same finger-nails, and in some the tail is lacking, while in none has it the prehensile character that it has in the sapajous and spider-monkeys. Intellectually, too, they rank higher, and in comparing them, in this respect, with a horse or a dog due allowance has to be made for an important element. Dogs and horses have been domesticated for thousands of years, so that we have to deal in their cases not only with the natural, but the inherited intelligence derived from generations of contact with the intellectual king of the animal world. In the case of the monkeys, at best we can have but the inheritance of but a very few generations of domesticated forms; everything else is the result of the animal's own advance unassisted by man. This allowance made, monkeys are seen to occupy a position next to man, if not above an Australian, a Bushman, or a Patagonian in everything but speech. Mimicry is a characteristic of the lower races of man, but in this point they are excelled by monkeys. Curiosity is shared by both, but the savage does not excel the monkey in the intelligent way in which he endeavors to find out the nature and meaning of things. Monkeys show both affection and sympathy for one another, and instances are cited where they have apparently attempted to reproach man for his

FIG. 536. — Mandrill (*Cynocephalus mormon*).

cruelty, while their sense of the ludicrous and their dislike of ridicule are well known. These, however, are more of the animal characteristics: they have intellectual ones as well. They possess a good memory and the power to associate together cause and effect, and to draw conclusions. They remember that in one instance certain results followed certain acts; and so when essentially similar conditions recur, they employ the same tactics and the same operations as before. In short, they reason, and in every way they seem to show that Mivart's famous dictum must be reversed: their mental powers differ from those of man in degree, not in kind. Man is not the only tool-using animal. Monkeys will use stones for hammers, understand the principle of the lever, can put in and take out screws, and do many things which we commonly think can be done by man alone. The monkey who threw a rope's end to a fellow that had fallen overboard acted from something else than mere instinct.

The baboons have a tail short or of moderate length, and a head so like that of a dog as to have given the genus the name *Cynocephalus*, which means dog-headed. They are disagreeable creatures, going in large troops, and making their homes in rocky fastnesses. They feed upon fruits and insects, and are more terrestrial than many of their relatives. Related to them are the drill and the mandrill, the latter without exception the ugliest of all apes, if not of all mammals. It is a familiar form in every menagerie, and the brilliant colors of the patches of naked skin on the sides of the nose and on the hindquarters, are sure to attract attention. Many frequently think that this ornamentation is largely due to the showman's brush: nothing of the sort; that brilliant red and that livid blue are nature's painting; those ridges upon the nose are but indices of the ugly, snarling temper of the beast. And yet there is not a doubt but that to mandrill eyes these phantasies of color, that swollen, furrowed nose, and that stub of a tail are the most beautiful things in nature, and that Mrs. Mandrill believes her huge lord to be as handsome as a picture.

In the oriental regions the macaques take the place of the African baboons; the species are numerous, but none is more common than the Javanese form, shown in our full-page plate. Some of the macaques are held in high esteem by the East Indians, even though they are like the rest, — arrant thieves and very destructive to the cultivated fields. They live well in confinement: indeed, they are the most hardy of all the apes, and hence they are frequently seen in our country in the peripatetic animal shows, as well as in the collections in the zoological gardens. Dr. Oswald has given a very pleasant account of one way in which these animals are captured, which follows in a condensed form.

When the British first effected a settlement at Singapore, the traffic in monkeys soon became a regular branch of industry. The ubiquitous Chinaman used to go on trapping expeditions to the hills at a time when the macaques were rather hard up for provisions, and could be baited with 'fuddle-cakes': *i.e.* rice-bread soaked in a mixture of sugar and rum. The trapper used to hide behind a tree, and let the monkey assemblage enjoy his bounty until their antics suggested that it was time for him to rush in. Experience soon taught the little mountaineers to change their

A GROUP OF JAVANESE MACAQUES (*Macacus cynomolgus*).

1–5, Orang.   ATTITUDES OF APES.   6–8, Gibbon.

tactics. Instead of devouring the fuddle-cakes on the spot, they learned to gather them up, and defer the feast till they reached a retreat where they could hope to be left alone in their glory. But the trappers, too, have since changed their plan. They manufacture a sort of narrow-necked jars, and, after filling them with a *mélange* of sirup and alcohol, they tie them firmly to the root of a tree, and withdraw out of sight. The monkeys come down and sip the nectar, a little at a time, till many a mickle has muddled their perceptions, to the degree which the founder of Buddhism would have called the first stage of Nirvana, — indifference to earthly concernments in general. The trapper then approaches, and collects his guests, whose exalted feelings often manifest themselves in a peculiar way. Some receive their captor with open arms ; some hug their bottles with approbative grunts ; while others lie on the ground, contemplating the sky in ecstatic silence. One of the macaques — the magot, or Barbary ape — deserves mention on account of its extension west over northern Africa and across the Strait to the Rock of Gibraltar. It is the only species of the whole monkey tribe living wild in all Europe, and this small colony contains the only representatives in that geographical division.

The doctrine of metempsychosis is a constituent of many an oriental religion, and if one believe that after death he is to be transformed into a monkey, he could hope for nothing better than the East Indians' faith, that their ancestors now live in the troops of hoonumauns which throng the region of the Ganges, and that they, too, are to be members of the family. In personal appearance these apes are not very handsome, but in dexterity they possibly exceed any of the other apes and monkeys. Listen to the way one writer describes them: "Without wings agility could hardly go farther ; from the standpoint of a practical anatomist it is almost inconceivable how muscles and sinews, apparently so very similar to our own, can execute such movements. Without the least visible effort, the marvellous half-bird darts through the air in a wide zig-zag, merely touching a branch here and there ; upward suddenly, with a series of mighty swings, regardless and apparently forgetful of obstacles ; down, with a gradationed swing that looks like a single leap; up again, with a flying rebound, through a tangle-work of branches, yet at the same time watching his comrades, aiming and parrying a slap, or dodging a shower of missiles ; then, with a sudden grab, a quick contraction of the hind legs, and the acrobat sits motionless on a projecting branch, watching a movement in the grass that has not escaped his eye during his headlong evolutions."

Of the man-like apes, the culmination of simian development, volumes might be written, so fraught with interest are these, our nearest brute relations. The gibbons and the orang-utan are oriental forms ; the chim-

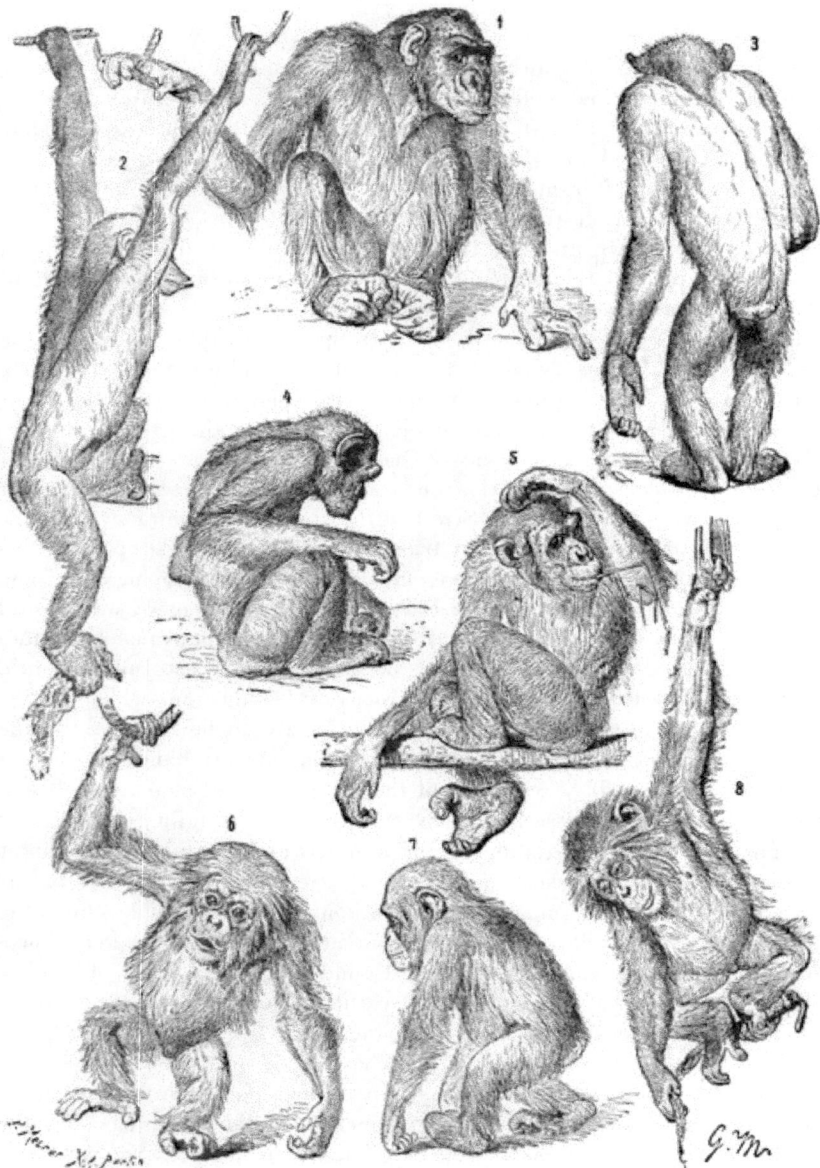

1-5, Tschego.   ATTITUDES OF APES.   6-8, Chimpanzee.

panzee and the gorilla are African. Of these the gibbons (there are several species) are less like man than the rest, and hence they possess less interest for us; but still a short extract from the pages of Mr. Forbes is so pathetic that the reader will pardon its introduction. "The wau-wau gibbons," he says, " have a wonderfully human look in their eyes; and it was with great distress that I witnessed the death of the only one I ever shot. Falling on its back with a dull thud on the ground, it raised itself on its elbows, passed its long, taper fingers over the wound, gave a woeful

FIG. 537. — Orang-utan (*Simia satyrus*).

look at them, and fell back at full length, dead — 'saperti orang' (just like a man), as my boy remarked."

For our knowledge of the habits of the orang-utan (the name is Malay, and means man-of-the-woods) we are largely indebted to the researches of A. R. Wallace, who spent a long time in Sumatra and Borneo, where this species occurs. A characteristic feature of these large apes is the great length of the arms. The Greek artists claimed that the outspread arms of a perfect man should equal the height of the body; but the largest orang obtained by Wallace, 'a giant,' was four feet and two inches tall, while

CHIMPANZEE (*Troglodytes niger*).

the arms spread to a distance of seven feet nine inches. The result is, that when standing erect, the hands nearly touch the ground. The orang-utan lives in the trees of the forest, making its way from limb to limb in a partially erect position, swinging himself easily from one tree to another by the aid of the interlacing branches. It never jumps, and never seems to hurry, and yet gets along about as fast as a man running on the ground beneath. It makes a nest of branches in the trees, in which it sleeps, not coming out until the sun has dried the dew from the leaves. It is remarkably strong, and the Dyaks informed Mr. Wallace that it killed the crocodile by main strength, pulling open its jaws and ripping up its throat.

Mr. Wallace obtained a young orang about a foot in length, which was hanging to its mother when she was shot. He kept it for about three months, and gives a full account of its habits. It was toothless when taken, and was fed with rice-water, no milk being obtainable. It soon came to enjoy its daily bath, and the subsequent wiping and combing of its long hair made it perfectly happy. In order to exercise its muscles, a short ladder was made, and on this the baby was allowed to hang for a quarter of an hour at a time. At first it seemed pleased, but as it could not get all its hands in a comfortable position, it would loose one after the other and drop to the floor. Sometimes when hanging with two hands it would loose one and cross it to the opposite shoulder, grasping its own hair, and as this seemed much more agreeable than the stick, it would attempt it with the other hand; of course tumbling to the floor, where it would cross both arms and lie perfectly contented. This fondness for fur led to the manufacture of an artificial mother out of skin. This answered for a short time, but then the baby tried to suck and succeeded in filling its mouth with hair, so this had to be given up. While it was kept it cut its two upper front teeth, but the lack of proper food probably led to the attack of intermittent fever which carried it off.

The chimpanzee and the gorilla are more closely related to each other than is either to the orang-utan. In having shorter arms in proportion to the body they are more like man, as they also are in the mode of articulation of the hip-joint and the number of bones in the wrist, while the orang, on the other hand, has the ribs (twelve) the same in number as in man, and the gorilla and chimpanzee have thirteen. In this way various correspondences and differences could be pointed out, but the total would indicate that the two forms last mentioned were nearer to man than is the orang. Similar comparisons between these two further show that the gorilla excels the chimpanzee in this respect.

The habits of the chimpanzee are much better known than those of the gorilla, for the reason that many have been kept in confinement; and

GORILLA (*Troglodytes gorilla*): male, female, and young.

in the zoological gardens of New York and Philadelphia no animals excite more interest than "Mr. Crowley" and his relatives in the City of Brotherly Love. According to Mr. Brown, the superintendent of the garden at Philadelphia, the specimens there show powers of reasoning — not instinct — of no mean order. They soon seemed to understand the effects of a mirror. They recognized the fact that a snake could not come through the glass front of the cage. And when one of the two died, its companion gave unmistakable signs of grief, nor was it deceived into thinking its own reflection in a mirror was its departed playmate. The chimpanzee recognizes its own superiority to other monkeys and its inferiority to man.

The chimpanzee ranges over central and western equatorial Africa. It reaches a height of about five feet. Like the orang-utan it is an arboreal animal and a fruit-eater. It does not make houses in the branches, but curls itself up and spends the night wherever it may be. It is not very afraid of man, and ravages the fields of the inhabitants, who for this reason destroy it whenever seen, even organizing hunting-parties to rid the neighborhood of the pests.

The gorilla is more retiring in its character, making its home in the densest forests of the western coast of equatorial Africa, its range being partially overlapped by that of the chimpanzee. Indeed, these two forms are said frequently to interbreed, and doubtless some, at least, of the many varieties of the man-like apes of Africa are due to such mixture of blood. The gorilla is the larger of the two, reaching a height of five feet seven or eight inches, while the arms are proportionately but little longer than the same members in man. The young gorilla, except for the hair and feet, has a very human aspect; but this decreases with growth, and this divergence is especially noticeable in the face on account of the retreating forehead and the development of strong, bony crests above the eyes.

To an American missionary, Dr. Savage, is due the first reliable account of these animals, but since his day many travelers have visited its haunts and given us accounts of its habits in its native wilds so far as it was possible to observe them; not a very easy task in the case of such a strong and ferocious beast. With all the accounts there is a considerable agreement, with one exception; Du Chaillu conflicts in many points with the rest, but which is correct on all points we must leave others to decide. It is, however, certain that all the statements contained in the pages of the author mentioned are not to be relied upon, no matter how interesting they may be. Gorillas try to escape from man; but when brought to bay, they will fight savagely, their strong arms and their savage jaws being the chief offensive and defensive organs. Gorillas are rather solitary in their habits;

and though they frequently take to the trees, they are far less arboreal than their cousin, the chimpanzee.

Last to be mentioned of all the animal kingdom is man, and to even outline his natural history would require volumes. While differing but little in structure from the great apes just mentioned, there is an intellectual superiority, which marks him off from all the rest, and places him on a higher plane. Between a gorilla or a chimpanzee and the most cultured of the Aryan race the differences are immense; but if these animals are compared with some of the lower races of man, the distance between the two is not so great, and, indeed, in some respects the apes do not suffer much by the comparison. The whole animal kingdom is a tree of which man is the highest twig, but the branch of apes reaches nearly as far upwards.

INDEX.